Solutions Manual
to Accompany

Introductory Physics

Rebekah L. Mays and John D. Mays

NOVARE
CLASSICAL ACADEMIC PRESS

Camp Hill, Pennsylvania
2021

Solutions Manual to Accompany Accelerated Studies in Physics and Chemistry

© Classical Academic Press®, 2021

Edition 2.1

ISBN: 978-0-9972845-5-3

Classical Academic Press

515 S. 32nd Street

Camp Hill, PA 17011

www.ClassicalAcademicPress.com/Novare/

IS.09.21

Contents

Acknowledgement

I (John) wish to express my gratitude to my daughter Rebekah Mays for carefully and meticulously compiling these solutions. Any errors that remain in this volume are my own responsibility.

Preface

This solutions manual contains fully detailed solutions for all of the computational problems contained in my text *Introductory Physics*. Teachers and students using that text should find this manual to be a valuable resource.

When comparing your results to the results shown here and to those in the text, keep in mind that the last digit is always uncertain because of the way significant digits in measurements are defined. When two results match except for a small difference in the most precise digit, we say that the results match. Because of rounding in calculators, it will not be uncommon for results shown here to differ from the answer key in the text or from your result by or two in the most precise digit.

I have checked and double checked the solutions to make them as accurate as possible. However, in any manual of this kind it is inevitable that errors remain. If you find an error, we would be much obliged if you would inform us of it by sending an email to info@novarescienceandmath.com.

Chapter 2

Unit Conversions

1.

$$1750 \text{ m} \cdot \frac{100 \text{ cm}}{1 \text{ m}} \cdot \frac{1 \text{ in}}{2.54 \text{ cm}} \cdot \frac{1 \text{ ft}}{12 \text{ in}} = 5740 \text{ ft}$$

2.

$$3.54 \text{ g} \cdot \frac{1 \text{ kg}}{1000 \text{ g}} = 0.00354 \text{ kg}$$

3.

$$41.11 \text{ mL} \cdot \frac{1 \text{ L}}{1000 \text{ mL}} = 0.04111 \text{ L}$$

4.

$$7 \times 10^8 \text{ m} \cdot \frac{100 \text{ cm}}{1 \text{ m}} \cdot \frac{1 \text{ in}}{2.54 \text{ cm}} \cdot \frac{1 \text{ ft}}{12 \text{ in}} \cdot \frac{1 \text{ mi}}{5280 \text{ ft}} = 4 \times 10^5 \text{ mi}$$

5.

$$1.5499 \times 10^{-12} \text{ mm} \cdot \frac{1 \text{ m}}{1000 \text{ mm}} = 1.5499 \times 10^{-15} \text{ m}$$

6.

$$750 \text{ cm}^3 \cdot \frac{1 \text{ mL}}{1 \text{ cm}^3} \cdot \frac{1 \text{ L}}{1000 \text{ mL}} \cdot \frac{1 \text{ m}^3}{1000 \text{ L}} = 7.5 \times 10^{-4} \text{ m}^3$$

7.

$$2.9979 \times 10^8 \frac{\text{m}}{\text{s}} \cdot \frac{100 \text{ cm}}{1 \text{ m}} \cdot \frac{1 \text{ in}}{2.54 \text{ cm}} \cdot \frac{1 \text{ ft}}{12 \text{ in}} = 9.8356 \times 10^8 \frac{\text{ft}}{\text{s}}$$

8.

$$168 \text{ hr} \cdot \frac{60 \text{ min}}{1 \text{ hr}} \cdot \frac{60 \text{ s}}{1 \text{ min}} = 605{,}000 \text{ s}$$

9.

$$5570 \frac{\text{kg}}{\text{m}^3} \cdot \frac{1000 \text{ g}}{1 \text{ kg}} \cdot \frac{1 \text{ m}^3}{1000 \text{ L}} \cdot \frac{1 \text{ L}}{1000 \text{ mL}} \cdot \frac{1 \text{ mL}}{1 \text{ cm}^3} = 5.57 \frac{\text{g}}{\text{cm}^3}$$

10.

$$45 \ \frac{\text{gal}}{\text{s}} \cdot \frac{3.786 \ \text{L}}{1 \ \text{gal}} \cdot \frac{1 \ \text{m}^3}{1000 \ \text{L}} \cdot \frac{60 \ \text{s}}{1 \ \text{min}} = 1.0 \times 10^1 \ \frac{\text{m}^3}{\text{min}}$$

11.

$$600,000 \ \frac{\text{ft}^3}{\text{s}} \cdot \frac{(0.3048 \ \text{m})^3}{1 \ \text{ft}^3} \cdot \frac{1000 \ \text{L}}{1 \ \text{m}^3} \cdot \frac{60 \ \text{s}}{1 \ \text{min}} \cdot \frac{60 \ \text{min}}{1 \ \text{hr}} = 6 \times 10^{10} \ \frac{\text{L}}{\text{hr}}$$

12.

$$5200 \ \text{mL} \cdot \frac{1 \ \text{L}}{1000 \ \text{mL}} \cdot \frac{1 \ \text{m}^3}{1000 \ \text{L}} = 5.2 \times 10^{-3} \ \text{m}^3$$

13.

$$5.65 \times 10^2 \ \text{mm}^2 \cdot \frac{1 \ \text{cm}}{10 \ \text{mm}} \cdot \frac{1 \ \text{cm}}{10 \ \text{mm}} \cdot \frac{1 \ \text{in}}{2.54 \ \text{cm}} \cdot \frac{1 \ \text{in}}{2.54 \ \text{cm}} = 0.876 \ \text{in}^2$$

14.

$$32.16 \ \frac{\text{ft}}{\text{s}^2} \cdot \frac{12 \ \text{in}}{1 \ \text{ft}} \cdot \frac{2.54 \ \text{cm}}{1 \ \text{in}} \cdot \frac{1 \ \text{m}}{100 \ \text{cm}} = 9.802 \ \frac{\text{m}}{\text{s}^2}$$

15.

$$5.001 \ \frac{\mu\text{g}}{\text{s}} \cdot \frac{1 \ \text{g}}{10^6 \ \mu\text{g}} \cdot \frac{1 \ \text{kg}}{1000 \ \text{g}} \cdot \frac{60 \ \text{s}}{1 \ \text{min}} = 3.001 \times 10^{-4} \ \frac{\text{kg}}{\text{min}}$$

16.

$$4.771 \ \frac{\text{g}}{\text{mL}} \cdot \frac{1 \ \text{kg}}{1000 \ \text{g}} \cdot \frac{1000 \ \text{mL}}{1 \ \text{L}} \cdot \frac{1000 \ \text{L}}{1 \ \text{m}^3} = 4771 \ \frac{\text{kg}}{\text{m}^3}$$

17.

$$13.6 \ \frac{\text{g}}{\text{cm}^3} \cdot \frac{1000 \ \text{mg}}{1 \ \text{g}} \cdot \frac{100 \ \text{cm}}{1 \ \text{m}} \cdot \frac{100 \ \text{cm}}{1 \ \text{m}} \cdot \frac{100 \ \text{cm}}{1 \ \text{m}} = 1.36 \times 10^{10} \ \frac{\text{mg}}{\text{m}^3}$$

18.

$$93,000,000 \ \text{mi} \cdot \frac{5280 \ \text{ft}}{1 \ \text{mi}} \cdot \frac{0.3048 \ \text{m}}{1 \ \text{ft}} \cdot \frac{100 \ \text{cm}}{1 \ \text{m}} = 1.5 \times 10^{13} \ \text{cm}$$

19.

$$65 \ \frac{\text{mi}}{\text{hr}} \cdot \frac{5280 \ \text{ft}}{1 \ \text{mi}} \cdot \frac{0.3048 \ \text{m}}{1 \ \text{ft}} \cdot \frac{1 \ \text{hr}}{60 \ \text{min}} \cdot \frac{1 \ \text{min}}{60 \ \text{s}} = 29 \ \frac{\text{m}}{\text{s}}$$

20.

$$633 \text{ nm} \cdot \frac{1 \text{ m}}{1 \times 10^9 \text{ nm}} \cdot \frac{100 \text{ cm}}{1 \text{ m}} \cdot \frac{1 \text{ in}}{2.54 \text{ cm}} = 2.49 \times 10^{-5} \text{ in}$$

21.

$$0.05015 \cdot 3.00 \times 10^8 \frac{\text{m}}{\text{s}} \cdot \frac{60 \text{ s}}{1 \text{ min}} \cdot \frac{60 \text{ min}}{1 \text{ hr}} \cdot \frac{1 \text{ ft}}{0.3048 \text{ m}} \cdot \frac{1 \text{ mi}}{5280 \text{ ft}} = 3.37 \times 10^7 \frac{\text{mi}}{\text{hr}}$$

Motion Exercises

1.

$$d = 25.1 \text{ mi} \cdot \frac{5280 \text{ ft}}{1 \text{ mi}} \cdot \frac{0.3048 \text{ m}}{1 \text{ ft}} = 4.04 \times 10^4 \text{ m}$$

$$t = 0.50 \text{ hr} \cdot \frac{60 \text{ min}}{1 \text{ hr}} \cdot \frac{60 \text{ s}}{1 \text{ min}} = 1800 \text{ s}$$

$$v = ?$$

$$v = \frac{d}{t} = \frac{4.04 \times 10^4 \text{ m}}{1800 \text{ s}} = 22 \frac{\text{m}}{\text{s}}$$

2.

$$22 \frac{\text{m}}{\text{s}} \cdot \frac{1 \text{ km}}{1000 \text{ m}} \cdot \frac{60 \text{ s}}{1 \text{ min}} \cdot \frac{60 \text{ min}}{1 \text{ hr}} = 79 \frac{\text{km}}{\text{hr}}$$

3.

$$t = 4.25 \text{ hr} \cdot \frac{3600 \text{ s}}{\text{hr}} = 15{,}300 \text{ s}$$

$$v = 5.0000 \frac{\text{km}}{\text{hr}} \cdot \frac{1000 \text{ m}}{\text{km}} \cdot \frac{1 \text{ hr}}{3600 \text{ s}} = 1.389 \frac{\text{m}}{\text{s}}$$

$$d = ?$$

$$v = \frac{d}{t}$$

$$d = vt$$

$$d = 1.389 \frac{\text{m}}{\text{s}} \cdot 15{,}300 \text{ s} = 21{,}300 \text{ m} \cdot \frac{1 \text{ km}}{1000 \text{ m}} = 21.3 \text{ km}$$

4.

$$21.3 \text{ km} \cdot \frac{1000 \text{ m}}{1 \text{ km}} \cdot \frac{1 \text{ ft}}{0.3048 \text{ m}} \cdot \frac{1 \text{ mi}}{5{,}280 \text{ ft}} = 13.2 \text{ mi}$$

5.

$$150.0 \; \frac{\text{mi}}{\text{hr}} \cdot \frac{5280 \text{ ft}}{1 \text{ mi}} \cdot \frac{0.3048 \text{ m}}{1 \text{ ft}} \cdot \frac{1 \text{ km}}{1000 \text{ m}} = 241.4 \; \frac{\text{km}}{\text{hr}}$$

6.

$$v = 150.0 \; \frac{\text{mi}}{\text{hr}} \cdot \frac{1609 \text{ m}}{\text{mi}} \cdot \frac{1 \text{ hr}}{3600 \text{ s}} = 67.04 \; \frac{\text{m}}{\text{s}}$$

$$d = 10.0 \text{ mi} \cdot \frac{1609 \text{ m}}{\text{mi}} = 16{,}090 \text{ m}$$

$$t = ?$$

$$v = \frac{d}{t}$$

$$t = \frac{d}{v} = \frac{16{,}090 \text{ m}}{67.04 \; \dfrac{\text{m}}{\text{s}}} = 240.0 \text{ s} \cdot \frac{1 \text{ min}}{60 \text{ s}} = 4.00 \text{ min}$$

7.

$$d = 3.0 \text{ km} \cdot \frac{1000 \text{ m}}{1 \text{ km}} = 3.0 \times 10^{3} \text{ m}$$

$$t = 1 \text{ hr } 20.0 \text{ min} = 80.0 \text{ min} \cdot \frac{60 \text{ s}}{1 \text{ min}} = 4.80 \times 10^{3} \text{ s}$$

$$v = ?$$

$$v = \frac{d}{t} = \frac{3.0 \times 10^{3} \text{ m}}{4.80 \times 10^{3} \text{ s}} = 0.63 \; \frac{\text{m}}{\text{s}}$$

8.

$$v_i = 0$$

$$v_f = 45 \; \frac{\text{mi}}{\text{hr}} \cdot \frac{1 \text{ hr}}{60 \text{ min}} \cdot \frac{1 \text{ min}}{60 \text{ s}} \cdot \frac{5{,}280 \text{ ft}}{1 \text{ mi}} \cdot \frac{0.3048 \text{ m}}{1 \text{ ft}} = 20.1 \; \frac{\text{m}}{\text{s}}$$

$$t = 36 \text{ s}$$

$$a = ?$$

$$a = \frac{v_f - v_i}{t} = \frac{20.1 \dfrac{\text{m}}{\text{s}} - 0}{36 \text{ s}} = 0.56 \; \frac{\text{m}}{\text{s}^{2}}$$

9.

$$v_i = 31 \, \frac{m}{s}$$

$$t = 17 \, s$$

$$v_f = 22 \, \frac{m}{s}$$

$$a = ?$$

$$a = \frac{v_f - v_i}{t} = \frac{22 \, \frac{m}{s} - 31 \, \frac{m}{s}}{17 \, s} = -0.53 \, \frac{m}{s^2}$$

10.

$$d = 14.5 \, m$$

$$v = c = 3.00 \times 10^8 \, \frac{m}{s}$$

$$t = ?$$

$$v = \frac{d}{t}$$

$$t = \frac{d}{v} = \frac{14.5 \, m}{3.00 \times 10^8 \, \frac{m}{s}} = 4.83 \times 10^{-8} \, s \cdot \frac{1 \times 10^9 \, ns}{s} = 48.3 \, ns$$

11.

$$v_i = 0$$

$$v_f = 0.80 \cdot 3.00 \times 10^8 \, \frac{m}{s} = 2.40 \times 10^8 \, \frac{m}{s}$$

$$t = 18 \, hr \, 6 \, min \, 45 \, s = 64{,}800 \, s + 360 \, s + 45 \, s = 65{,}205 \, s$$

$$a = ?$$

$$a = \frac{v_f - v_i}{t} = \frac{2.40 \times 10^8 \, \frac{m}{s} - 0}{65{,}205 \, s} = 3680 \, \frac{m}{s^2}$$

12.

$$d = 8.96 \times 10^9 \text{ km} \cdot \frac{1000 \text{ m}}{1 \text{ km}} = 8.96 \times 10^{12} \text{ m}$$

$$v = 3.45 \times 10^5 \ \frac{\text{m}}{\text{s}}$$

$$t = ?$$

$$v = \frac{d}{t}$$

$$t = \frac{d}{v} = \frac{8.96 \times 10^{12} \text{ m}}{3.45 \times 10^5 \ \dfrac{\text{m}}{\text{s}}} = 2.597 \times 10^7 \text{ s} \cdot \frac{1 \text{ hr}}{3600 \text{ s}} \cdot \frac{1 \text{ day}}{24 \text{ hr}} = 301 \text{ days}$$

13.

$$a = 5.556 \times 10^6 \ \frac{\text{cm}}{\text{s}^2} \cdot \frac{1 \text{ m}}{100 \text{ cm}} = 5.556 \times 10^4 \ \frac{\text{m}}{\text{s}^2}$$

$$t = 45 \text{ ms} \cdot \frac{1 \text{ s}}{1000 \text{ ms}} = 4.5 \times 10^{-2} \text{ s}$$

$$v_i = 0$$

$$v_f = ?$$

$$a = \frac{v_f - v_i}{t}$$

$$v_f = at + v_i = (5.556 \times 10^4 \ \frac{\text{m}}{\text{s}^2})(4.5 \times 10^{-2} \text{ s}) + (0 \ \frac{\text{m}}{\text{s}}) = 2.5 \times 10^3 \ \frac{\text{m}}{\text{s}}$$

14.

$$v_i = 4.005 \times 10^3 \ \frac{\text{m}}{\text{s}}$$

$$a = 23.1 \ \frac{\text{m}}{\text{s}^2}$$

$$t = 13.5 \text{ s}$$

$$v_f = ?$$

$$a = \frac{v_f - v_i}{t}$$

$$v_f = at + v_i = (23.1 \ \frac{\text{m}}{\text{s}^2} \cdot 13.5 \text{ s}) + 4.005 \times 10^3 \ \frac{\text{m}}{\text{s}} = 4.32 \times 10^3 \ \frac{\text{m}}{\text{s}}$$

15.

$$v = c = 2.9979 \times 10^8 \ \frac{m}{s}$$

$$d = 1.4965 \times 10^8 \ km \cdot \frac{1000 \ m}{1 \ km} = 1.4965 \times 10^{11} \ m$$

$$t = ?$$

$$v = \frac{d}{t}$$

$$t = \frac{d}{v} = \frac{1.4965 \times 10^{11} \ m}{2.9979 \times 10^8 \ \frac{m}{s}} = 499.18 \ s \cdot \frac{1 \ min}{60 \ s} = 8.3197 \ min$$

Chapter 3

Newton's Second Law Practice Problems

1.

$m = 1880 \text{ kg}$

$a = 1.50 \ \dfrac{\text{m}}{\text{s}^2}$

$F = ?$

$a = \dfrac{F}{m}$

$F = ma = 1880 \text{ kg} \cdot 1.50 \ \dfrac{\text{m}}{\text{s}^2} = 2820 \text{ N}$

2.

$m = 188.4 \text{ g} \cdot \dfrac{1 \text{ kg}}{1000 \text{ g}} = 0.1884 \text{ kg}$

$g = 9.80 \ \dfrac{\text{m}}{\text{s}^2}$

$F_w = ?$

$F_w = 0.1884 \text{ kg} \cdot 9.80 \ \dfrac{\text{m}}{\text{s}^2} = 1.85 \text{ N}$

3.

$F = 250.0 \text{ N}$

$m = 144{,}000 \text{ mg} \cdot \dfrac{1 \text{ g}}{1000 \text{ mg}} \cdot \dfrac{1 \text{ kg}}{1000 \text{ g}} = 0.144 \text{ kg}$

$a = ?$

$a = \dfrac{F}{m} = \dfrac{250.0 \text{ N}}{0.144 \text{ kg}} = 1740 \ \dfrac{\text{m}}{\text{s}^2}$

4.

$$a = 2.3 \; \frac{m}{s^2}$$

$$F = 230{,}000 \text{ N}$$

$$m = ?$$

$$a = \frac{F}{m}$$

$$m = \frac{F}{a} = \frac{230{,}000 \text{ N}}{2.3 \; \dfrac{m}{s^2}} = 1.0 \times 10^5 \text{ kg}$$

5.

$$a = 0.0022 \; \frac{mi}{hr^2} \cdot \frac{5280 \text{ ft}}{1 \text{ mi}} \cdot \frac{0.3048 \text{ m}}{1 \text{ ft}} \cdot \frac{1 \text{ hr}}{3600 \text{ s}} \cdot \frac{1 \text{ hr}}{3600 \text{ s}} = 2.732 \times 10^{-7} \; \frac{m}{s^2}$$

$$m = 2.2 \text{ Mg} \cdot \frac{1 \times 10^6 \text{ g}}{1 \text{ Mg}} \cdot \frac{1 \text{ kg}}{1000 \text{ g}} = 2.2 \times 10^3 \text{ kg}$$

$$F = ?$$

$$a = \frac{F}{m}$$

$$F = ma = 2.2 \times 10^3 \text{ kg} \cdot 2.732 \times 10^{-7} \; \frac{m}{s^2} = 6.0 \times 10^{-4} \text{ N}$$

6.

$$F_w = 125.1 \text{ lb} \cdot \frac{4.45 \text{ N}}{1 \text{ lb}} = 556.7 \text{ N}$$

$$g = 9.80 \; \frac{m}{s^2}$$

$$m = ?$$

$$F_w = mg$$

$$m = \frac{F_w}{g} = \frac{556.7 \text{ N}}{9.80 \; \dfrac{m}{s^2}} = 56.8 \text{ kg}$$

7.

$m = 56.8$ kg

$F_w = 20.9$ lb $\cdot \dfrac{4.45\ \text{N}}{1\ \text{lb}} = 93.01$ N

$g_m = ?$

$F_w = mg_m$

$g_m = \dfrac{F_w}{m} = \dfrac{93.01\ \text{N}}{56.8\ \text{kg}} = 1.64\ \dfrac{\text{m}}{\text{s}^2}$

8.

$v_i = 0$

$v_f = 125.0\ \dfrac{\text{m}}{\text{s}}$

$t = 22.00$ ms $\cdot \dfrac{1\ \text{s}}{1000\ \text{ms}} = 2.200 \times 10^{-2}$ s

$F = 142.0$ N

$m = ?$

$a = \dfrac{v_f - v_i}{t} = \dfrac{125.0\ \dfrac{\text{m}}{\text{s}} - 0}{2.200 \times 10^{-2}\ \text{s}} = 5681.8\ \dfrac{\text{m}}{\text{s}^2}$

$a = \dfrac{F}{m}$

$m = \dfrac{F}{a} = \dfrac{142.0\ \text{N}}{5681.8\ \dfrac{\text{m}}{\text{s}^2}} = 0.024992\ \text{kg} \cdot \dfrac{1000\ \text{g}}{1\ \text{kg}} = 24.99$ g

---

9.

$m = 4.5 \text{ kg}$

$v_i = 0$

$v_f = 8.00 \ \dfrac{\text{mi}}{\text{hr}} \cdot \dfrac{1 \text{ hr}}{3600 \text{ s}} \cdot \dfrac{5280 \text{ ft}}{1 \text{ mi}} \cdot \dfrac{0.3048 \text{ m}}{1 \text{ ft}} = 3.6 \ \dfrac{\text{m}}{\text{s}}$

$t = 500 \text{ ms} \cdot \dfrac{1 \text{ s}}{1000 \text{ ms}} = 0.5 \text{ s}$

$F = ?$

$a = \dfrac{v_f - v_i}{t} = \dfrac{3.6 \ \dfrac{\text{m}}{\text{s}} - 0}{0.5 \text{ s}} = 7.2 \ \dfrac{\text{m}}{\text{s}^2}$

$a = \dfrac{F}{m}$

$F = ma = 4.5 \text{ kg} \cdot 7.2 \ \dfrac{\text{m}}{\text{s}^2} = 30 \text{ N}$

10.

$v_i = 2500.0 \ \dfrac{\text{km}}{\text{hr}} \cdot \dfrac{1 \text{ hr}}{3600 \text{ s}} \cdot \dfrac{1000 \text{ m}}{1 \text{ km}} = 694.4 \ \dfrac{\text{m}}{\text{s}}$

$t = 8.000 \text{ s}$

$F = 45,450 \text{ N}$

$v_f = 2750 \ \dfrac{\text{km}}{\text{hr}} \cdot \dfrac{1000 \text{ m}}{1 \text{ km}} \cdot \dfrac{1 \text{ hr}}{3600 \text{ s}} = 763.9 \ \dfrac{\text{m}}{\text{s}}$

$m = ?$

$a = \dfrac{v_f - v_i}{t} = \dfrac{763.9 \ \dfrac{\text{m}}{\text{s}} - 694.4 \ \dfrac{\text{m}}{\text{s}}}{8.000 \text{ s}} = 8.688 \ \dfrac{\text{m}}{\text{s}^2}$

$a = \dfrac{F}{m}$

$m = \dfrac{F}{a} = \dfrac{45,450 \text{ N}}{8.688 \ \dfrac{\text{m}}{\text{s}^2}} = 5230 \text{ kg}$

11.

$$m = 166 \text{ g} \cdot \frac{1 \text{ kg}}{1000 \text{ g}} = 0.166 \text{ kg}$$

$$F = 0.0450 \text{ N}$$

$$v_i = 0$$

$$t = 2.1 \text{ s}$$

$$v_f = ?$$

$$a = \frac{F}{m} = \frac{0.0450 \text{ N}}{0.166 \text{ kg}} = 0.2711 \; \frac{\text{m}}{\text{s}^2}$$

$$a = \frac{v_f - v_i}{t}$$

$$v_f = at + v_i = \left(0.2711 \; \frac{\text{m}}{\text{s}^2} \cdot 2.1 \text{ s}\right) + 0 = 0.57 \; \frac{\text{m}}{\text{s}}$$

12.

$$m = 1.673 \times 10^{-18} \; \mu\text{g} \cdot \frac{1 \text{ g}}{1 \times 10^6 \; \mu\text{g}} \cdot \frac{1 \text{ kg}}{1000 \text{ g}} = 1.673 \times 10^{-27} \text{ kg}$$

$$v_i = 0$$

$$v_f = c \cdot 0.0005 = 3.00 \times 10^8 \cdot 0.0005 = 1.50 \times 10^5 \; \frac{\text{m}}{\text{s}}$$

$$t = 455 \text{ ns} \cdot \frac{1 \text{ s}}{1 \times 10^9 \text{ ns}} = 4.55 \times 10^{-7} \text{ s}$$

$$F = ?$$

$$a = \frac{v_f - v_i}{t} = \frac{1.50 \times 10^5 \; \frac{\text{m}}{\text{s}} - 0}{4.55 \times 10^{-7} \text{ s}} = 3.30 \times 10^{11} \; \frac{\text{m}}{\text{s}^2}$$

$$a = \frac{F}{m}$$

$$F = ma = 1.673 \times 10^{-27} \text{ kg} \cdot 3.30 \times 10^{11} \; \frac{\text{m}}{\text{s}^2} = 5.52 \times 10^{-16} \text{ N} \cdot \frac{1 \text{ GN}}{1 \times 10^9 \text{ N}} = 5.52 \times 10^{-25} \text{ GN}$$

13.

$$m = 6.548 \text{ Gg} \cdot \frac{1 \times 10^9 \text{ g}}{1 \text{ Gg}} \cdot \frac{1 \text{ kg}}{1000 \text{ g}} = 6.548 \times 10^6 \text{ kg}$$

$$v_i = 8.35 \frac{\text{mi}}{\text{hr}} \cdot \frac{1609 \text{ m}}{1 \text{ mi}} \cdot \frac{1 \text{ hr}}{3600 \text{ s}} = 3.732 \frac{\text{m}}{\text{s}}$$

$$v_f = 0$$

$$t = 0.288 \text{ min} \cdot \frac{60 \text{ s}}{1 \text{ min}} = 17.28 \text{ s}$$

$$F = ?$$

$$a = \frac{v_f - v_i}{t} = \frac{0 - 3.732 \frac{\text{m}}{\text{s}}}{17.28 \text{ s}} = -0.216 \frac{\text{m}}{\text{s}^2}$$

$$a = \frac{F}{m}$$

$$F = ma$$

$$F = 6.548 \times 10^6 \text{ kg} \cdot \left(-0.216 \frac{\text{m}}{\text{s}^2} \right) = -1.41 \times 10^6 \text{ N}$$ (The negative sign indicates

that the force is a stopping force, i.e., it opposes the motion.)

14.

$$v_i = 3.5 \frac{\text{cm}}{\text{s}} \cdot \frac{1 \text{ m}}{100 \text{ cm}} = 0.035 \frac{\text{m}}{\text{s}}$$

$$v_f = 18.5 \frac{\text{cm}}{\text{s}} \cdot \frac{1 \text{ m}}{100 \text{ cm}} = 0.185 \frac{\text{m}}{\text{s}}$$

$$t = 220 \text{ ms} \cdot \frac{1 \text{ s}}{1000 \text{ ms}} = 0.22 \text{ s}$$

$$a = ?$$

$$a = \frac{v_f - v_i}{t} = \frac{0.185 \frac{\text{m}}{\text{s}} - 0.035 \frac{\text{m}}{\text{s}}}{0.22 \text{ s}} = 0.68 \frac{\text{m}}{\text{s}^2}$$

15.

$m = 45{,}500 \text{ kg}$

$v_i = 0 \; \dfrac{\text{m}}{\text{s}}$

$v_f = 55 \; \dfrac{\text{m}}{\text{s}}$

$t = 6.4 \text{ s}$

$a = ?$

$F = ?$

$$a = \frac{v_f - v_i}{t} = \frac{55 \; \dfrac{\text{m}}{\text{s}} - 0}{6.4 \text{ s}} = 8.59 \; \frac{\text{m}}{\text{s}^2}$$

$$a = \boxed{8.6 \; \frac{\text{m}}{\text{s}^2}}$$

$a = \dfrac{F}{m}$

$$F = ma = 45{,}500 \text{ kg} \cdot 8.59 \; \frac{\text{m}}{\text{s}^2} = 3.9 \times 10^5 \text{ N}$$

16.

$m = 8.5 \text{ g} \cdot \dfrac{1 \text{ kg}}{1000 \text{ g}} = 0.0085 \text{ kg}$

$a = 18{,}500 \; \dfrac{\text{m}}{\text{s}^2}$

$F = ?$

$a = \dfrac{F}{m}$

$$F = ma = 0.0085 \text{ kg} \cdot 18{,}500 \; \frac{\text{m}}{\text{s}^2} = 160 \text{ N}$$

Chapter 4

Classroom Energy Computation Examples

1.

$m = 1.00 \times 10^5$ kg

$h = 240 \text{ ft} \cdot \dfrac{0.3048 \text{ m}}{1 \text{ ft}} = 73.15 \text{ m}$

$g = 9.80 \dfrac{\text{m}}{\text{s}^2}$

$E_G = ?$

$E_G = mgh = 1.00 \times 10^5 \text{ kg} \cdot 9.80 \dfrac{\text{m}}{\text{s}^2} \cdot 73.15 \text{ m} = 72{,}000{,}000 \text{ J}$

2.

$m = 25 \text{ g} \cdot \dfrac{1 \text{ kg}}{1000 \text{ g}} = 0.025 \text{ kg}$

$v = 556 \dfrac{\text{ft}}{\text{s}} \cdot \dfrac{0.3048 \text{ m}}{1 \text{ ft}} = 169.5 \dfrac{\text{m}}{\text{s}}$

$E_K = ?$

$E_K = \dfrac{1}{2}mv^2 = \dfrac{1}{2} \cdot 0.025 \text{ kg} \cdot \left(169.5 \dfrac{\text{m}}{\text{s}} \right)^2 = 360 \text{ J}$

3.

$d = 75 \text{ cm} \cdot \dfrac{1 \text{ m}}{100 \text{ cm}} = 0.75 \text{ m}$

$m = 12{,}500 \text{ g} \cdot \dfrac{1 \text{ kg}}{1000 \text{ g}} = 12.5 \text{ kg}$

$W = ?$

$W = Fd$

$F_w = mg = 12.5 \text{ kg} \cdot 9.80 \dfrac{\text{m}}{\text{s}^2} = 122.5 \text{ kg}$

$W = 122.5 \text{ kg} \cdot 0.75 \text{ m} = 92 \text{ J}$

4.

$m = 12.5 \text{ kg}$

$h = 0.75 \text{ m}$

$E_G = mgh = 0.75 \text{ m} \cdot 9.80 \ \dfrac{\text{m}}{\text{s}^2} \cdot 12.5 \text{ kg} = 92 \text{ J}$

5.

$m = 12.5 \text{ kg}$

$h_i = 0.75 \text{ m}$

$h_f = 0$

$v_i = 0$

$v_f = ?$

$E_{Gi} + E_{Ki} = E_{Gf} + E_{Kf}$

$E_{Kf} = E_{Gi} + E_{Ki} - E_{Gf} = 92 \text{ J} + 0 - 0 = 92 \text{ J}$

$v_f = \sqrt{\dfrac{2E_{Kf}}{m}} = \sqrt{\dfrac{2 \cdot 92 \text{ J}}{12.5 \text{ kg}}} = 3.8 \ \dfrac{\text{m}}{\text{s}}$

6.a.

$m = 255.8 \text{ g} \cdot \dfrac{1 \text{ kg}}{1000 \text{ g}} = 0.2558 \text{ kg}$

$h_i = 10.4 \text{ ft} \cdot \dfrac{0.3048 \text{ m}}{1 \text{ ft}} = 3.1699 \text{ m}$

$E_{Gi} = ?$

$E_{Gi} = mgh_i = 0.2558 \text{ kg} \cdot 9.80 \ \dfrac{\text{m}}{\text{s}^2} \cdot 3.1699 \text{ m} = 7.95 \text{ J}$

$h_f = 0$

$v_i = 0$

$v_f = ?$

$E_{Gi} + E_{Ki} = E_{Gf} + E_{Kf}$

$E_{Ki} = 0$

$E_{Gf} = 0$

$E_{Kf} = E_{Gi} + E_{Ki} - E_{Gf} = 7.95 \text{ J} + 0 \text{ J} - 0 \text{ J} = 7.95 \text{ J}$

$v_f = \sqrt{\dfrac{2E_{Kf}}{m}} = \sqrt{\dfrac{2 \cdot 7.95 \text{ J}}{0.2558 \text{ kg}}} = 7.88 \ \dfrac{\text{m}}{\text{s}}$

Energy Calculations Set 1

1.

$m = 1.31 \times 10^3 \text{ kg}$

$h = 177.44 \text{ ft} \cdot \dfrac{0.3048 \text{ m}}{1 \text{ ft}} = 54.084 \text{ m}$

$E_G = ?$

$E_G = mgh = 1.31 \times 10^3 \text{ kg} \cdot 9.80 \ \dfrac{\text{m}}{\text{s}^2} \cdot 54.084 \text{ m} = 694{,}000 \text{ J}$

2.

$m = 2345 \text{ kg}$

$v = 31 \ \dfrac{\text{mi}}{\text{hr}} \cdot \dfrac{5280 \text{ ft}}{1 \text{ mi}} \cdot \dfrac{0.3048 \text{ m}}{1 \text{ ft}} \cdot \dfrac{1 \text{ hr}}{3600 \text{ s}} = 13.858 \ \dfrac{\text{m}}{\text{s}}$

$E_K = \tfrac{1}{2}mv^2 = 0.5 \cdot 2345 \text{ kg} \cdot \left(13.858 \ \dfrac{\text{m}}{\text{s}} \right)^2 = 230{,}000 \text{ J}$

3.

$$d = 61.7 \text{ cm} \cdot \frac{1 \text{ m}}{100 \text{ cm}} = 0.617 \text{ m}$$

$$m = 17.5 \text{ kg}$$

$$W = ?$$

$$a = \frac{F}{m}$$

$$F_w = mg = 17.5 \text{ kg} \cdot 9.80 \, \frac{\text{m}}{\text{s}^2} = 171.5 \text{ N}$$

$$W = F_w d = 171.5 \text{ N} \cdot 0.617 \text{ m} = 106 \text{ J}$$

4.

$$h = 61.7 \text{ cm} \cdot \frac{1 \text{ m}}{100 \text{ cm}} = 0.617 \text{ m}$$

$$m = 17.5 \text{ kg}$$

$$E_G = ?$$

$$E_G = mgh = 17.5 \text{ kg} \cdot 9.80 \, \frac{\text{m}}{\text{s}^2} \cdot 0.617 \text{ m} = 106 \text{ J}$$

5.

$$m = 17.5 \text{ kg}$$

$$h_f = 0.617 \text{ m}$$

$$h_i = 0$$

$$v_i = 0$$

$$v_f = ?$$

$$E_{Gi} + E_{Ki} = E_{Gf} + E_{Kf}$$

$$E_{Kf} = E_{Gi} + E_{Ki} - E_{Gf} = 106 \text{ J} + 0 \text{ J} - 0 \text{ J} = 106 \text{ J}$$

$$v_f = \sqrt{\frac{2E_{Kf}}{m}} = \sqrt{\frac{2 \cdot 106 \text{ J}}{17.5 \text{ kg}}} = 12.1 \, \frac{\text{m}}{\text{s}} = 3.48 \, \frac{\text{m}}{\text{s}}$$

6.

$$m = 122 \text{ g} \cdot \frac{1 \text{ kg}}{1000 \text{ g}} = 0.122 \text{ kg}$$

$$v_i = 13.75 \, \frac{\text{m}}{\text{s}}$$

$$v_f = 0$$

$$h_i = 0$$

$$h_f = ?$$

$$E_{Ki} = \tfrac{1}{2}mv^2 = 0.5 \cdot 0.122 \text{ kg} \cdot \left(13.75 \, \frac{\text{m}}{\text{s}}\right)^2 = 11.53 \text{ J}$$

$$E_{Gi} + E_{Ki} = E_{Gf} + E_{Kf}$$

$$E_{Gf} = E_{Gi} + E_{Ki} - E_{Kf} = 0 + 11.53 \text{ J} - 0 = 11.53 \text{ J}$$

$$E_{Gf} = mgh_f$$

$$h_f = \frac{E_{Gf}}{mg} = \frac{11.53 \text{ J}}{0.122 \text{ kg} \cdot 9.8 \, \frac{\text{m}}{\text{s}^2}} = 9.65 \text{ m}$$

7.

$$m = 325 \text{ g} \cdot \frac{1 \text{ kg}}{1000 \text{ g}} = 0.325 \text{ kg}$$

$$h_i = 36.1 \text{ m}$$

$$h_f = 0$$

$$v_i = 0$$

$$v_f = ?$$

$$E_{Gi} = mgh_i = 0.325 \text{ kg} \cdot 9.80 \, \frac{\text{m}}{\text{s}^2} \cdot 36.1 \text{ m} = 114.98 \text{ J}$$

$$E_{Gi} + E_{Ki} = E_{Gf} + E_{Kf}$$

$$E_{Kf} = E_{Gi} + E_{Ki} - E_{Gf} = 114.98 \text{ J} + 0 - 0$$

$$E_{Kf} = 114.98 \text{ J}$$

$$E_{Kf} = \frac{1}{2}mv^2$$

$$v_f = \sqrt{\frac{2E_{Kf}}{m}} = \sqrt{\frac{2 \cdot 114.98 \text{ J}}{0.325 \text{ kg}}} = 26.6 \, \frac{\text{m}}{\text{s}}$$

8.

$F = 735$ N

$d = 26$ m

$W = ?$

$W = Fd = 735$ N $\cdot 26$ m $= 19,000$ J

Energy Calculations Set 2

1.a.

$F_w = 20 \cdot 80.0 \text{ lb} \cdot \dfrac{4.45 \text{ N}}{1 \text{ lb}} = 7120$ N

$h = 8.5$ m

$g = 9.80 \ \dfrac{\text{m}}{\text{s}^2}$

$m = ?$

$F_w = mg$

$m = \dfrac{F_w}{g}$

$m = \dfrac{7120 \text{ N}}{9.80 \ \dfrac{\text{m}}{\text{s}^2}} = 727$ kg

---

1.b

$F = 7120$ N

$d = 8.5$ m

$W = Fd = 7120$ N $\cdot 8.5$ m

$W = 6.0 \times 10^4$ J

1.c.

$m = 727$ kg

$g = 9.80 \ \dfrac{\text{m}}{\text{s}^2}$

$h = 8.5$ m

$E_G = mgh$

$E_G = 727 \text{ kg} \cdot 9.80 \ \dfrac{\text{m}}{\text{s}^2} \cdot 8.5 \text{ m} = 6.0 \times 10^4$ J

1.d.i.

$m = 727 \text{ kg}$

$v_i = 0$

$E_{Ki} = ?$

$E_{Ki} = \dfrac{1}{2}mv_i^2 = \dfrac{1}{2} \cdot 727 \text{ kg} \cdot 0^2 = 0 \text{ J}$

1.d.ii.

$E_{Gi} = 6.0 \times 10^4 \text{ J}$

$E_{Gf} = 0$

$E_{Kf} = ?$

$E_{Gi} + E_{Ki} = E_{Gf} + E_{Kf}$

$E_{Kf} = E_{Gi} + E_{Ki} - E_{Gf} = 6.0 \times 10^4 \text{ J} + 0 - 0 = 6.0 \times 10^4 \text{ J}$

1.e.

$m = 727 \text{ kg}$

$E_{Kf} = 6.0 \times 10^4 \text{ J}$

$v_f = ?$

$E_{Kf} = \frac{1}{2}mv_f^2$

$v_f = \sqrt{\dfrac{2E_{Kf}}{m}} = \sqrt{\dfrac{2 \cdot 6.0 \times 10^4 \text{ J}}{727 \text{ kg}}} = 13 \ \dfrac{\text{m}}{\text{s}}$

2.a.

$$F_w = 3{,}193 \text{ lb} \cdot \frac{4.45 \text{ N}}{1 \text{ lb}} = 14{,}209 \text{ N}$$

$$m = ?$$

$$F_w = mg$$

$$m = \frac{F_w}{g}$$

$$m = \frac{14{,}209 \text{ N}}{9.80 \, \frac{\text{m}}{\text{s}^2}} = 1450 \text{ kg}$$

$$h = 16 \text{ m}$$

$$E_G = mgh = 1450 \text{ kg} \cdot 9.80 \, \frac{\text{m}}{\text{s}^2} \cdot 16 \text{ m} = 227{,}360 \text{ J}$$

$$E_G = 230{,}000 \text{ J}$$

2.c.

$$E_{Kf} = ?$$

$$E_{G_i} + E_{Ki} = E_{Gf} + E_{Kf}$$

$$E_{Kf} = E_{G_i} + E_{Ki} - E_{Gf}$$

$$E_{Kf} = 230{,}000 \text{ J} + 0 - 0 = 230{,}000 \text{ J}$$

2.d.

$$E_{Kf} = 230{,}000 \text{ J}$$

$$m = 1450 \text{ kg}$$

$$v_f = ?$$

$$v_f = \sqrt{\frac{2E_{Kf}}{m}} = \sqrt{\frac{2 \cdot 230{,}000 \text{ J}}{1450 \text{ kg}}} = 18 \, \frac{\text{m}}{\text{s}}$$

Energy Calculations Set 3

1.a.

$$F_w = 27.05 \text{ lb} \cdot \frac{4.45 \text{ N}}{1 \text{ lb}} = 120.37 \text{ N}$$

$$d = 185 \text{ cm} \cdot \frac{1 \text{ m}}{100 \text{ cm}} = 1.85 \text{ m}$$

$$W = Fd = 120.37 \text{ N} \cdot 1.85 \text{ m} = 223 \text{ J}$$

1.b

$F_w = 120.37 \text{ N}$

$h = 1.85 \text{ m}$

$g = 9.80 \, \dfrac{\text{m}}{\text{s}^2}$

$m = ?$

$F_w = mg$

$m = \dfrac{F_w}{g}$

$m = \dfrac{120.37 \text{ N}}{9.80 \, \dfrac{\text{m}}{\text{s}^2}} = 12.28 \text{ kg}$

$E_G = ?$

$E_G = mgh = 12.28 \text{ kg} \cdot 9.80 \, \dfrac{\text{m}}{\text{s}^2} \cdot 1.85 \text{ m} = 223 \text{ J}$

1.c.

$m = 12.28 \text{ kg}$

$g = 9.80 \, \dfrac{\text{m}}{\text{s}^2}$

$h = 0 \text{ m}$

$E_G = ?$

$E_G = mgh = 12.28 \text{ kg} \cdot 9.80 \, \dfrac{\text{m}}{\text{s}^2} \cdot 0 \text{ m} = 0 \text{ J}$

1.d.

$E_{Gi} = 222.7 \text{ J}$

$E_{Gf} = 0 \text{ J}$

$E_{Ki} = 0 \text{ J}$

$E_{Kf} = ?$

$E_{Gi} + E_{Ki} = E_{Gf} + E_{Kf}$

$E_{Kf} = E_{Gi} + E_{Ki} - E_{Gf} = 222.7 \text{ J} + 0 - 0 = 223 \text{ J}$

1.e.

$E_{Kf} = 222.7 \text{ J}$

$m = 12.28 \text{ kg}$

$v_f = ?$

$$v_f = \sqrt{\frac{2E_{Kf}}{m}} = \sqrt{\frac{2 \cdot 222.7 \text{ J}}{12.28 \text{ kg}}} = 6.02 \; \frac{\text{m}}{\text{s}}$$

2.

$$d = 197 \text{ ft} \cdot \frac{0.3048 \text{ m}}{1 \text{ ft}} = 60.05 \text{ m}$$

$m = 6.016 \times 10^6 \text{ kg}$

$$F_w = mg = 6.016 \times 10^6 \text{ kg} \cdot 9.80 \; \frac{\text{m}}{\text{s}^2} = 5.896 \times 10^7 \text{ N}$$

$W = ?$

$W = F_w d = 5.896 \times 10^7 \text{ N} \cdot 60.05 \text{ m}$

$$W = 3.54 \times 10^9 \text{ J} \cdot \frac{1 \text{ GJ}}{1 \times 10^9 \text{ J}} = 3.54 \text{ GJ}$$

3.

$m = 5{,}122 \text{ kg}$

$h_i = 25.0 \text{ m}$

$v_f = ?$

$$E_{Gi} = mgh_i = 5{,}122 \text{ kg} \cdot 9.80 \; \frac{\text{m}}{\text{s}^2} \cdot 25.0 \text{ m} = 1.2549 \times 10^6 \text{ J}$$

$E_{Gi} = E_{Kf} = 1.2549 \times 10^6 \text{ J}$

$$v_f = \sqrt{\frac{2E_{Kf}}{m}} = \sqrt{\frac{2 \cdot 1.2549 \times 10^6 \text{ J}}{5{,}122 \text{ kg}}} = 22.1 \; \frac{\text{m}}{\text{s}}$$

4.a.

$$F_w = 104.6 \text{ lb} \cdot \frac{4.45 \text{ N}}{1 \text{ lb}} = 4.6547 \times 10^2 \text{ N}$$

$$d = 13 \text{ steps} \cdot \frac{16.5 \text{ cm}}{1 \text{ step}} \cdot \frac{1 \text{ m}}{100 \text{ cm}} = 2.145 \text{ m}$$

$$W = ?$$

$$W = F_w d = 4.6547 \times 10^2 \text{ N} \cdot 2.145 \text{ m} = 998 \text{ J}$$

4.b.

$$F_w = 4.6547 \times 10^2 \text{ N}$$

$$F_w = mg$$

$$m = \frac{F_w}{g} = \frac{4.6547 \times 10^2 \text{ N}}{9.80 \frac{\text{m}}{\text{s}^2}} = 47.497 \text{ kg}$$

$$h_i = 2.145 \text{ m}$$

$$v_f = ?$$

$$E_{Gi} = mgh_i = 47.497 \text{ kg} \cdot 9.80 \frac{\text{m}}{\text{s}^2} \cdot 2.145 \text{ m}$$

$$E_{Gi} = 998.43 \text{ J}$$

$$E_{Gi} + E_{Ki} = E_{Gf} + E_{Kf}$$

$$E_{Kf} = E_{Gi} + E_{Ki} - E_{Gf} = 998.43 \text{ J} + 0 - 0$$

$$E_{Kf} = 998.43 \text{ J}$$

$$v_f = \sqrt{\frac{2 E_{Kf}}{m}} = \sqrt{\frac{2 \cdot 998.43 \text{ J}}{47.497 \text{ kg}}} = 6.48 \frac{\text{m}}{\text{s}}$$

5.

$$m = 351 \text{ g} \cdot \frac{1 \text{ kg}}{1000 \text{ g}} = 0.351 \text{ kg}$$

$$v_i = 500.00 \ \frac{\text{cm}}{\text{s}} \cdot \frac{1 \text{ m}}{100 \text{ cm}} = 5.00 \ \frac{\text{m}}{\text{s}}$$

$$v_f = 0 \ \frac{\text{m}}{\text{s}}$$

$$h_i = 0 \text{ m}$$

$$h_f = ?$$

$$E_{Ki} = \tfrac{1}{2}mv^2 = 0.5 \cdot 0.351 \text{ kg} \cdot \left(5.00 \ \frac{\text{m}}{\text{s}} \right)^2$$

$$E_{Ki} = 4.388 \text{ J}$$

$$E_{Gi} + E_{Ki} = E_{Gf} + E_{Kf}$$

$$E_{Gf} = E_{Gi} + E_{Ki} - E_{Kf} = 0 \text{ J} + 4.388 \text{ J} - 0 \text{ J}$$

$$E_{Gf} = 4.388 \text{ J}$$

$$E_{Gf} = mgh_f$$

$$h_f = \frac{E_{Gf}}{mg} = \frac{4.388 \text{ J}}{0.351 \text{ kg} \cdot 9.80 \ \dfrac{\text{m}}{\text{s}^2}} = 1.28 \text{ m}$$

Chapter 5

Basic Momentum Computations

1.

$m = 1350$ kg

$v = 35 \dfrac{\text{mi}}{\text{hr}} \cdot \dfrac{5{,}280 \text{ ft}}{\text{mi}} \cdot \dfrac{0.3048 \text{ m}}{1 \text{ ft}} \cdot \dfrac{1 \text{ hr}}{3{,}600 \text{ s}} = 15.6 \dfrac{\text{m}}{\text{s}}$

$p = ?$

$p = mv = 1350 \text{ kg} \cdot 15.6 \dfrac{\text{m}}{\text{s}} = 21{,}000 \dfrac{\text{kg} \cdot \text{m}}{\text{s}}$

2.

$m = 8.71 \times 10^{-4} \text{ g} \cdot \dfrac{1 \text{ kg}}{1000 \text{ g}} = 8.71 \times 10^{-7} \text{ kg}$

$v = 725.5 \dfrac{\text{m}}{\text{s}}$

$p = ?$

$p = mv = 8.71 \times 10^{-7} \text{ kg} \cdot 725.5 \dfrac{\text{m}}{\text{s}} = 6.32 \times 10^{-4} \dfrac{\text{kg} \cdot \text{m}}{\text{s}}$

3.

$m = 144.3 \text{ g} \cdot \dfrac{1 \text{ kg}}{1000 \text{ g}} = 0.1443 \text{ kg}$

$v = 99.55 \dfrac{\text{mi}}{\text{hr}} \cdot \dfrac{5{,}280 \text{ ft}}{1 \text{ mi}} \cdot \dfrac{0.3048 \text{ m}}{1 \text{ ft}} \cdot \dfrac{1 \text{ hr}}{3{,}600 \text{ s}} = 44.50 \dfrac{\text{m}}{\text{s}}$

$p = ?$

$p = mv = 0.1443 \text{ kg} \cdot 44.503 \dfrac{\text{m}}{\text{s}} = 6.421 \dfrac{\text{kg} \cdot \text{m}}{\text{s}}$

4.

$$v = 1{,}847 \; \frac{\text{ft}}{\text{s}} \cdot \frac{0.3048 \text{ m}}{1 \text{ ft}} = 5.630 \times 10^2 \; \frac{\text{m}}{\text{s}}$$

$$p = 25{,}565 \; \frac{\text{kg} \cdot \text{m}}{\text{s}}$$

$$m = ?$$

$$F_w = ?$$

$$p = mv$$

$$m = \frac{p}{v} = \frac{25{,}565 \; \dfrac{\text{kg} \cdot \text{m}}{\text{s}}}{5.630 \times 10^2 \; \dfrac{\text{m}}{\text{s}}} = 45.41 \text{ kg}$$

$$F_w = mg = 45.41 \text{ kg} \cdot 9.80 \; \frac{\text{m}}{\text{s}^2} = 445.02 \text{ N} \cdot \frac{1 \text{ lb}}{4.45 \text{ N}} = 1.00 \times 10^2 \text{ lb}$$

5.

$$m = 9.11 \times 10^{-28} \text{ g} \cdot \frac{1 \text{ kg}}{1000 \text{ g}} = 9.11 \times 10^{-31} \text{ kg}$$

$$v = 0.3375 \cdot 3.00 \times 10^8 \; \frac{\text{m}}{\text{s}} = 1.0125 \times 10^8 \; \frac{\text{m}}{\text{s}}$$

$$p = ?$$

$$p = mv = 9.11 \times 10^{-31} \text{ kg} \cdot 1.0125 \times 10^8 \; \frac{\text{m}}{\text{s}} = 9.22 \times 10^{-23} \; \frac{\text{kg} \cdot \text{m}}{\text{s}}$$

6.

$$m = 1.673 \times 10^{-24} \text{ g} \cdot \frac{1 \text{ kg}}{1000 \text{ g}} = 1.673 \times 10^{-27} \text{ kg}$$

$$p = 4.5516 \times 10^{-23} \; \frac{\text{kg} \cdot \text{m}}{\text{s}}$$

$$v = ?$$

$$p = mv$$

$$v = \frac{p}{m} = \frac{4.5516 \times 10^{-23} \; \dfrac{\text{kg} \cdot \text{m}}{\text{s}}}{1.673 \times 10^{-27} \text{ kg}} = 2.721 \times 10^4 \; \frac{\text{m}}{\text{s}}$$

$$2.721 \times 10^4 \; \frac{\text{m}}{\text{s}} \cdot \frac{100 \text{ cm}}{1 \text{ m}} \cdot \frac{1 \text{ in}}{2.54 \text{ cm}} = 1.071 \times 10^6 \; \frac{\text{in}}{\text{s}}$$

7.

$$F_w = 47{,}055 \text{ tons} \cdot \frac{2{,}000 \text{ lb}}{1 \text{ ton}} \cdot \frac{4.45 \text{ N}}{1 \text{ lb}} = 4.1879 \times 10^8 \text{ N}$$

$$v = 55.75 \frac{\text{mi}}{\text{hr}} \cdot \frac{5{,}280 \text{ ft}}{1 \text{ mi}} \cdot \frac{0.3048 \text{ m}}{1 \text{ ft}} \cdot \frac{1 \text{ hr}}{3{,}600 \text{ s}} = 24.92 \frac{\text{m}}{\text{s}}$$

$$p = ?$$

$$F_w = mg$$

$$m = \frac{F_w}{g} = \frac{4.1879 \times 10^8 \text{ N}}{9.80 \frac{\text{m}}{\text{s}^2}} = 4.27 \times 10^7 \text{ kg}$$

$$p = mv = 4.27 \times 10^7 \text{ kg} \cdot 24.92 \frac{\text{m}}{\text{s}} = 1{,}060{,}000{,}000 \frac{\text{kg} \cdot \text{m}}{\text{s}}$$

8.

$$m = 5.98 \times 10^{24} \text{ kg}$$

$$l = 5.84 \times 10^8 \text{ mi} \cdot \frac{5{,}280 \text{ ft}}{1 \text{ mi}} \cdot \frac{0.3048 \text{ m}}{1 \text{ ft}} = 9.399 \times 10^{11} \text{ m}$$

$$p = ?$$

$$t = \frac{365 \text{ days}}{1 \text{ orbit}} \cdot \frac{24 \text{ hr}}{1 \text{ day}} \cdot \frac{3{,}600 \text{ s}}{1 \text{ hr}} = 3.15 \times 10^7 \text{ s}$$

$$v = \frac{9.399 \times 10^{11} \text{ m}}{3.15 \times 10^7 \text{ s}} = 2.98 \times 10^4 \frac{\text{m}}{\text{s}}$$

$$p = mv = 5.98 \times 10^{24} \text{ kg} \cdot 2.98 \times 10^4 \frac{\text{m}}{\text{s}} = 1.78 \times 10^{29} \frac{\text{kg} \cdot \text{m}}{\text{s}}$$

9.

$$v = 13.25 \frac{\text{in}}{\text{s}} \cdot \frac{2.54 \text{ cm}}{1 \text{ in}} \cdot \frac{1 \text{ m}}{100 \text{ cm}} = 0.3366 \frac{\text{m}}{\text{s}}$$

$$p = 5.9805 \times 10^{-4} \frac{\text{kg} \cdot \text{m}}{\text{s}}$$

$$m = ?$$

$$p = mv$$

$$m = \frac{p}{v} = \frac{5.9805 \times 10^{-4} \frac{\text{kg} \cdot \text{m}}{\text{s}}}{0.3366 \frac{\text{m}}{\text{s}}} = 0.001777 \text{ kg} \cdot \frac{1000 \text{ g}}{1 \text{ kg}} = 1.777 \text{ g}$$

10.

$$p = 2{,}177.9 \ \frac{\text{kg} \cdot \text{m}}{\text{s}}$$

$$F_w = 188.5 \ \text{lb} \cdot \frac{4.45 \ \text{N}}{1 \ \text{lb}} = 838.83 \ \text{N}$$

$$v = ?$$

$$F_w = mg$$

$$m = \frac{F_w}{g} = \frac{838.83 \ \text{N}}{9.80 \ \dfrac{\text{m}}{\text{s}^2}} = 85.594 \ \text{kg}$$

$$p = mv$$

$$v = \frac{p}{m} = \frac{2{,}177.9 \ \dfrac{\text{kg} \cdot \text{m}}{\text{s}}}{85.594 \ \text{kg}} = 25.4 \ \frac{\text{m}}{\text{s}}$$

Conservation of Momentum Computations

1.

After the collision, the first proton has stopped moving, and the second is moving in the same direction as the first proton at 8,672 km/s.

2.

$$v_{1i} = 2.00 \ \frac{\text{mi}}{\text{hr}} \cdot \frac{5,280 \ \text{ft}}{1 \ \text{mi}} \cdot \frac{0.3048 \ \text{m}}{1 \ \text{ft}} \cdot \frac{1 \ \text{hr}}{60 \ \text{min}} \cdot \frac{1 \ \text{min}}{60 \ \text{s}} = 0.894 \ \frac{\text{m}}{\text{s}}$$

$$v_{1f} = 1.00 \ \frac{\text{mi}}{\text{hr}} \cdot \frac{5,280 \ \text{ft}}{1 \ \text{mi}} \cdot \frac{0.3048 \ \text{m}}{1 \ \text{ft}} \cdot \frac{1 \ \text{hr}}{60 \ \text{min}} \cdot \frac{1 \ \text{min}}{60 \ \text{s}} = 0.447 \ \frac{\text{m}}{\text{s}}$$

$$m_1 = 3,450 \ \text{kg}$$

$$m_2 = 1,150 \ \text{kg}$$

$$v_{2f} = ?$$

$$p_{1i} = p_{1f} + p_{2f}$$

$$p_{2f} = p_{1i} - p_{1f}$$

$$p_{1i} = m_1 v_{1i} = 3,450 \ \text{kg} \cdot 0.894 \ \frac{\text{m}}{\text{s}} = 3,084 \ \frac{\text{kg} \cdot \text{m}}{\text{s}}$$

$$p_{1f} = m_1 v_{1f} = 3,450 \ \text{kg} \cdot 0.447 \ \frac{\text{m}}{\text{s}} = 1,542 \ \frac{\text{kg} \cdot \text{m}}{\text{s}}$$

$$p_{2f} = 3,084 \ \frac{\text{kg} \cdot \text{m}}{\text{s}} - 1,542 \ \frac{\text{kg} \cdot \text{m}}{\text{s}} = 1,542 \ \frac{\text{kg} \cdot \text{m}}{\text{s}}$$

$$v_{2f} = \frac{p_{2f}}{m_2} = \frac{1,542 \ \frac{\text{kg} \cdot \text{m}}{\text{s}}}{1,150 \ \text{kg}} = 1.34 \ \frac{\text{m}}{\text{s}} \cdot \frac{3,600 \ \text{s}}{1 \ \text{hr}} \cdot \frac{1 \ \text{ft}}{0.3048 \ \text{m}} \cdot \frac{1 \ \text{mi}}{5,280 \ \text{ft}} = 3.00 \ \frac{\text{mi}}{\text{hr}}$$

3.

$m_1 = 3,450 \text{ kg}$

$m_2 = 1,150 \text{ kg}$

$v_{1i} = 0.89408 \dfrac{\text{m}}{\text{s}}$

$v_{2i} = 0 \dfrac{\text{m}}{\text{s}}$

$v_{1f} = 0.447 \dfrac{\text{m}}{\text{s}}$

$v_{2f} = 1.34 \dfrac{\text{m}}{\text{s}}$

$E_{Ki} = ?$

$E_{Kf} = ?$

$E_{Ki} = \dfrac{1}{2}m_1 v_{1i}{}^2 = \dfrac{1}{2} \cdot 3,450 \text{ kg} \cdot (0.894 \dfrac{\text{m}}{\text{s}})^2 = 1,380 \text{ J}$

$E_{Kf} = \dfrac{1}{2}m_1 v_{1f}{}^2 + \dfrac{1}{2}m_2 v_{2f}{}^2 = \left(\dfrac{1}{2} \cdot 3,450 \text{ kg} \cdot \left(0.447 \dfrac{\text{m}}{\text{s}} \right)^2 \right) + \left(\dfrac{1}{2} \cdot 1,150 \text{ kg} \cdot \left(1.34 \dfrac{\text{m}}{\text{s}} \right)^2 \right)$

$E_{Kf} = 1,380 \text{ J}$

$E_{Ki} = E_{Kf} = 1,380 \text{ J}$

4.

$$m_1 = 375.00 \text{ g} \cdot \frac{1 \text{ kg}}{1000 \text{ g}} = 0.375 \text{ kg}$$

$$m_2 = 5 \cdot 0.375 \text{ kg} = 1.875 \text{ kg}$$

$$v_{1f} = 0.7500 \ \frac{\text{m}}{\text{s}}$$

$$v_{2f} = 0.3750 \ \frac{\text{m}}{\text{s}}$$

$$v_{1i} = ?$$

$$p_{1i} = (-p_{1f}) + p_{2f}$$

$$p_{1f} = m_1 v_{1f} = 0.375 \text{ kg} \cdot 0.7500 \ \frac{\text{m}}{\text{s}} = 0.2813 \ \frac{\text{kg} \cdot \text{m}}{\text{s}}$$

$$p_{2f} = m_2 v_{2f} = 1.875 \text{ kg} \cdot 0.3750 \ \frac{\text{m}}{\text{s}} = 0.7031 \ \frac{\text{kg} \cdot \text{m}}{\text{s}}$$

$$p_{1i} = 0.7031 \ \frac{\text{kg} \cdot \text{m}}{\text{s}} - 0.2813 \ \frac{\text{kg} \cdot \text{m}}{\text{s}} = 0.4218 \ \frac{\text{kg} \cdot \text{m}}{\text{s}}$$

$$p_{1i} = m_1 v_{1i}$$

$$v_{1i} = \frac{p_{1i}}{m_1} = \frac{0.4218 \ \frac{\text{kg} \cdot \text{m}}{\text{s}}}{0.375 \text{ kg}} = 1.125 \ \frac{\text{m}}{\text{s}}$$

5.

Since the masses are equal, the moving hockey puck will come to rest, and the second hockey puck will start moving in the direction of the first puck at 9.75 m/s.

6.

$m_1 = 5.50 \text{ kg}$

$m_2 = 550.0 \text{ g} \cdot \dfrac{1 \text{ kg}}{1000 \text{ g}} = 0.5500 \text{ kg}$

$v_{1f} = 3.2727 \dfrac{\text{cm}}{\text{s}} \cdot \dfrac{1 \text{ m}}{100 \text{ cm}} = 0.032727 \dfrac{\text{m}}{\text{s}}$

$v_{2f} = 7.2727 \dfrac{\text{cm}}{\text{s}} \cdot \dfrac{1 \text{ m}}{100 \text{ cm}} = 0.072727 \dfrac{\text{m}}{\text{s}}$

$v_{1i} = ?$

$p_{1i} = p_{1f} + p_{2f}$

$p_{1f} = m_1 v_{1f} = 5.50 \text{ kg} \cdot 0.032727 \dfrac{\text{m}}{\text{s}} = 0.18000 \dfrac{\text{kg} \cdot \text{m}}{\text{s}}$

$p_{2f} = m_2 v_{2f} = 0.5500 \text{ kg} \cdot 0.072727 \dfrac{\text{m}}{\text{s}} = 0.040000 \dfrac{\text{kg} \cdot \text{m}}{\text{s}}$

$p_{1i} = 0.18000 \dfrac{\text{kg} \cdot \text{m}}{\text{s}} + 0.040000 \dfrac{\text{kg} \cdot \text{m}}{\text{s}} = 0.22000 \dfrac{\text{kg} \cdot \text{m}}{\text{s}}$

$p_{1i} = m_1 v_{1i}$

$v_{1i} = \dfrac{p_{1i}}{m_1} = \dfrac{0.22000 \dfrac{\text{kg} \cdot \text{m}}{\text{s}}}{5.50 \text{ kg}} = 0.0400 \dfrac{\text{m}}{\text{s}} \cdot \dfrac{100 \text{ cm}}{1 \text{ m}} = 4.00 \dfrac{\text{cm}}{\text{s}}$

7.

$m_1 = 5.50 \text{ kg}$

$m_2 = 0.5500 \text{ kg}$

$v_{1i} = 0.0400 \dfrac{\text{m}}{\text{s}}$

$v_{2i} = 0 \text{ J}$

$v_{1f} = 0.032727 \dfrac{\text{m}}{\text{s}}$

$v_{2f} = 0.072727 \dfrac{\text{m}}{\text{s}}$

$E_{Ki} = ?$

$E_{Kf} = ?$

$$E_{Ki} = \frac{1}{2} m_1 v_{1i}^{\,2} = \frac{1}{2} \cdot 5.50 \text{ kg} \cdot \left(0.0400 \ \frac{\text{m}}{\text{s}} \right)^2 = 0.00440 \text{ J}$$

$$E_{Kf} = \frac{1}{2} m_1 v_{1f}^{\,2} + \frac{1}{2} m_2 v_{2f}^{\,2} = \left(\frac{1}{2} \cdot 5.50 \text{ kg} \cdot \left(0.032727 \ \frac{\text{m}}{\text{s}} \right)^2 \right)$$

$$+ \left(\frac{1}{2} \cdot 0.5500 \text{ kg} \cdot \left(0.072727 \ \frac{\text{m}}{\text{s}} \right)^2 \right) = 0.00440 \text{ J}$$

$$E_{Ki} = E_{Kf} = 0.00440 \text{ J}$$

8.

$$m_1 = 144.3 \text{ g} \cdot \frac{1 \text{ kg}}{1000 \text{ g}} = 0.1443 \text{ kg}$$

$$m_2 = 10 \cdot 0.1443 \text{ kg} = 1.443 \text{ kg}$$

$$v_{1i} = 65.00 \frac{\text{mi}}{\text{hr}} \cdot \frac{5,280 \text{ ft}}{1 \text{ mi}} \cdot \frac{0.3048 \text{ m}}{1 \text{ ft}} \frac{1 \text{ hr}}{3,600 \text{ s}} = 29.06 \frac{\text{m}}{\text{s}}$$

$$v_{1f} = 53.1818 \frac{\text{mi}}{\text{hr}} \cdot \frac{5,280 \text{ ft}}{1 \text{ mi}} \cdot \frac{0.3048 \text{ m}}{1 \text{ ft}} \frac{1 \text{ hr}}{3,600 \text{ s}} = 23.7744 \frac{\text{m}}{\text{s}}$$

$$v_{2f} = ?$$

$$p_{1i} = (-p_{1f}) + p_{2f}$$

$$p_{2f} = p_{1i} + p_{1f}$$

$$p_{1f} = m_1 v_{1f} = 0.1443 \text{ kg} \cdot 23.7744 \frac{\text{m}}{\text{s}} = 3.4306 \frac{\text{kg} \cdot \text{m}}{\text{s}}$$

$$p_{1i} = m_1 v_{1i} = 0.1443 \text{ kg} \cdot 29.06 \frac{\text{m}}{\text{s}} = 4.1934 \frac{\text{kg} \cdot \text{m}}{\text{s}}$$

$$p_{2f} = 4.1934 \frac{\text{kg} \cdot \text{m}}{\text{s}} + 3.4306 \frac{\text{kg} \cdot \text{m}}{\text{s}} = 7.624 \frac{\text{kg} \cdot \text{m}}{\text{s}}$$

$$p_{2f} = m_2 v_{2f}$$

$$v_{2f} = \frac{p_{2f}}{m_2} = \frac{7.624 \frac{\text{kg} \cdot \text{m}}{\text{s}}}{1.443 \text{ kg}} = 5.2834 \frac{\text{m}}{\text{s}} \cdot \frac{1 \text{ ft}}{0.3048 \text{ m}} \cdot \frac{1 \text{ mi}}{5,280 \text{ ft}} \cdot \frac{3,600 \text{ s}}{1 \text{ hr}} = 11.82 \frac{\text{mi}}{\text{hr}}$$

Chapter 6

Volume, Mass, and Weight Exercises

1.

$$98.34 \ \frac{kg}{m^3} \cdot \frac{1000 \ g}{1 \ kg} \cdot \left(\frac{1 \ m}{100 \ cm} \right)^3 = 0.09834 \ \frac{g}{cm^3}$$

2.

$$42 \ mL \cdot \frac{1 \ L}{1000 \ mL} \cdot \frac{1 \ gal}{3.785 \ L} = 0.011 \ gal$$

3.

$$F_w = 18.5 \ lb \cdot \frac{4.45 \ N}{1 \ lb} = 82.3 \ N$$

$$m = ?$$

$$F_w = mg$$

$$m = \frac{F_w}{g} = \frac{82.3 \ N}{9.80 \ \frac{m}{s^2}} = 8.40 \ kg$$

4.

$$3.6711 \times 10^4 \ \frac{g}{mL} \cdot \frac{1 \ kg}{1000 \ g} \cdot \frac{1000 \ mL}{1 \ L} \cdot \frac{1000 \ L}{1 \ m^3} = 3.6711 \times 10^7 \ \frac{kg}{m^3}$$

5.

$$1.957 \times 10^4 \ in^3 \cdot \left(\frac{2.54 \ cm}{1 \ in} \right)^3 = 320,700 \ cm^3$$

6.

$$455 \ mL \cdot \frac{1 \ L}{1000 \ mL} \cdot \frac{1 \ m^3}{1000 \ L} = 0.000455 \ m^3$$

7.

$m = 46{,}000 \text{ kg}$

$F_w = ?$

$F_w = mg = 46{,}000 \text{ kg} \cdot 9.80 \, \dfrac{\text{m}}{\text{s}^2} = 450{,}000 \text{ N}$

$4.508 \times 10^5 \text{ N} \cdot \dfrac{1 \text{ lb}}{4.45 \text{ N}} = 1.0 \times 10^5 \text{ lb}$

8.

$32.11 \text{ L} \cdot \dfrac{1000 \text{ cm}^3}{1 \text{ L}} \cdot \left(\dfrac{1 \text{ in}}{2.54 \text{ cm}} \right)^3 = 1959 \text{ in}^3$

9.

$F_w = 14.89 \text{ N} \cdot \dfrac{1 \text{ lb}}{4.45 \text{ N}} = 3.35 \text{ lb}$

$m = ?$

$m = \dfrac{F_w}{g} = \dfrac{14.89 \text{ N}}{9.80 \, \dfrac{\text{m}}{\text{s}^2}} = 1.52 \text{ kg}$

10.

$36.00 \text{ cm}^3 \cdot \left(\dfrac{1 \text{ m}}{100 \text{ cm}} \right)^3 = 3.6 \times 10^{-5} \text{ m}^3$

11.

$9.11 \text{ m}^3 \cdot \left(\dfrac{100 \text{ cm}}{1 \text{ m}} \right)^3 = 9.11 \times 10^6 \text{ cm}^3$

12.

$4.11 \times 10^5 \text{ m}^3 \cdot \dfrac{1000 \text{ L}}{1 \text{ m}^3} = 4.11 \times 10^8 \text{ L}$

13.

$F_w = 55{,}789 \text{ lb} \cdot \dfrac{4.45 \text{ N}}{1 \text{ lb}} = 2.48261 \times 10^5 \text{ N}$

$m = \dfrac{F_w}{g} = \dfrac{2.48261 \times 10^5 \text{ N}}{9.80 \, \dfrac{\text{m}}{\text{s}^2}} = 25{,}300 \text{ kg}$

14.

$$5.022 \, \frac{\text{g}}{\text{cm}^3} \cdot \frac{1 \text{ kg}}{1000 \text{ g}} \cdot \left(\frac{100 \text{ cm}}{1 \text{ m}} \right)^3 = 5022 \, \frac{\text{kg}}{\text{m}^3}$$

15.

$$F_w = 50,000 \text{ N} \cdot \frac{1 \text{ lb}}{4.45 \text{ N}} = 10,000 \text{ lb}$$

$$m = \frac{F_w}{g} = \frac{50,000 \text{ N}}{9.80 \, \frac{\text{m}}{\text{s}^2}} = 5000 \text{ kg}$$

16.

$$1.75 \times 10^{-6} \text{ m}^3 \cdot \left(\frac{100 \text{ cm}}{1 \text{ m}} \right)^3 = 1.75 \text{ cm}^3$$

17.

$$100.5 \text{ ft}^3 \cdot \left(\frac{12 \text{ in}}{1 \text{ ft}} \right)^3 \cdot \left(\frac{2.54 \text{ cm}}{1 \text{ in}} \right)^3 \cdot \left(\frac{1 \text{ m}}{100 \text{ cm}} \right)^3 = 2.846 \text{ m}^3$$

18.

$$37 \text{ m}^3 \cdot \left(\frac{100 \text{ cm}}{1 \text{ m}} \right)^3 \cdot \left(\frac{1 \text{ in}}{2.54 \text{ cm}} \right)^3 = 2,300,000 \text{ in}^3$$

19.

$$750 \text{ cm}^3 \cdot \frac{1 \text{ L}}{1000 \text{ cm}^3} = 0.75 \text{ L}$$

20.

$$5,755,000 \text{ gal} \cdot \frac{3.786 \text{ L}}{1 \text{ gal}} \cdot \frac{1000 \text{ cm}^3}{1 \text{ L}} \cdot \left(\frac{1 \text{ m}}{100 \text{ cm}} \right)^3 = 21,790 \text{ m}^3$$

Density Exercises

1.

$m = 0.196$ g

$V = 100.1$ mL

$\rho = ?$

$$\rho = \frac{m}{V} = \frac{0.196 \text{ g}}{100.1 \text{ mL}} = 1.96 \times 10^{-3} \frac{\text{g}}{\text{mL}}$$

2.

$$\rho = 955 \frac{\text{kg}}{\text{m}^3} \cdot \frac{1000 \text{ g}}{1 \text{ kg}} \cdot \left(\frac{1 \text{ m}}{100 \text{ cm}}\right)^3 \cdot \frac{100 \text{ cm}^3}{1 \text{ L}} \cdot \frac{1 \text{ L}}{1000 \text{ mL}} = 0.955 \frac{\text{g}}{\text{mL}}$$

$m = 550$ g

$V = ?$

$$\rho = \frac{m}{V}$$

$$V = \frac{m}{\rho} = \frac{550 \text{ g}}{0.955 \frac{\text{g}}{\text{mL}}} = 580 \text{ mL}$$

3.

$$m = 15.7 \text{ kg} \cdot \frac{1000 \text{ g}}{1 \text{ kg}} = 1.57 \times 10^4 \text{ g}$$

$$\rho = 5.32 \frac{\text{g}}{\text{cm}^3}$$

$V = ?$

$$\rho = \frac{m}{V}$$

$$V = \frac{m}{\rho} = \frac{1.57 \times 10^4 \text{ g}}{5.32 \frac{\text{g}}{\text{cm}^3}} = 2{,}950 \text{ cm}^3$$

$$2.95 \times 10^3 \text{ cm}^3 \cdot \left(\frac{1 \text{ m}}{100 \text{ cm}}\right)^3 = 0.00295 \text{ m}^3$$

4.

$l = 3.00$ cm

$w = 3.00$ cm

$h = 3.00$ cm

$F_w = 5.336 \times 10^{-2}$ lb $\cdot \dfrac{4.45 \text{ N}}{1 \text{ lb}} = 2.375 \times 10^{-1}$ N

$\rho = ?$

$F_w = mg$

$m = \dfrac{F_w}{g} = \dfrac{2.375 \times 10^{-1} \text{ N}}{9.80 \dfrac{\text{m}}{\text{s}^2}} = 2.423 \times 10^{-2}$ kg $\cdot \dfrac{1000 \text{ g}}{1 \text{ kg}} = 24.23$ g

$V = l \cdot w \cdot h = (3.00 \text{ cm})^3 = 27 \text{ cm}^3$

$\rho = \dfrac{m}{V} = \dfrac{24.23 \text{ g}}{27 \text{ cm}^3} = 0.897 \ \dfrac{\text{g}}{\text{cm}^3}$

5.

$V_1 = 23.35$ mL

$V_2 = 27.79$ mL

$m = 32.1$ g

$\rho = ?$

$V_{Total} = 27.79 \text{ mL} - 23.35 \text{ mL} = 4.44 \text{ mL} \cdot \dfrac{1 \text{ mL}}{1 \text{ cm}^3} = 4.44 \text{ cm}^3$

$\rho = \dfrac{32.1 \text{ g}}{4.44 \text{ cm}^3} = 7.23 \ \dfrac{\text{g}}{\text{cm}^3}$

6.

Five cubic centimeters of gold have a greater mass than 5 cubic centimeters of silver, since its density is higher.

7.

The ratios g/cm³ and kg/L are equivalent, so all we have to do is divide each mass by its volume and compare.

3.9 g/5 mL = 0.78 g/mL

6.6 kg/7.5 L = 0.88 kg/L = 0.88 g/mL

Benzene is more dense.

8.

$$\rho = 7{,}830 \; \frac{\text{kg}}{\text{m}^3} \cdot \frac{1000 \text{ g}}{1 \text{ kg}} \cdot \left(\frac{1 \text{ m}}{100 \text{ cm}} \right)^3 = 7.83 \; \frac{\text{g}}{\text{cm}^3}$$

$l = 2.1$ cm

$w = 3.5$ cm

$m = 94.5$ g

$h = ?$

$$\rho = \frac{m}{V}$$

$$V = \frac{m}{\rho} = \frac{94.5 \text{ g}}{7.83 \; \dfrac{\text{g}}{\text{cm}^3}} = 12.07 \text{ cm}^3$$

$$V = l \cdot w \cdot h$$

$$h = \frac{V}{l \cdot w} = \frac{12.07 \text{ cm}^3}{2.1 \text{ cm} \cdot 3.5 \text{ cm}} = 1.6 \text{ cm}$$

9.

$$m = 306 \text{ g} \cdot \frac{1 \text{ kg}}{1000 \text{ g}} = 0.306 \text{ kg}$$

$$V = 22.5 \text{ mL} \cdot \frac{1 \text{ cm}^3}{1 \text{ mL}} \cdot \left(\frac{1 \text{ m}}{100 \text{ cm}} \right)^3 = 2.25 \times 10^{-5} \text{ m}^3$$

$\rho = ?$

$$\rho = \frac{m}{V} = \frac{0.306 \text{ kg}}{2.25 \times 10^{-5} \text{ m}^3} = 13{,}600 \; \frac{\text{kg}}{\text{m}^3}$$

10.

$$\rho = 1{,}000.0 \; \frac{\text{kg}}{\text{m}^3}$$

$$V = 5.6 \text{ L} \cdot \frac{1000 \text{ cm}^3}{1 \text{ L}} \cdot \left(\frac{1 \text{ m}}{100 \text{ cm}} \right)^3 = 5.6 \times 10^{-3} \text{ m}^3$$

$m = ?$

$$\rho = \frac{m}{V}$$

$$m = V\rho = 5.6 \times 10^{-3} \text{ m}^3 \cdot 1{,}000.0 \; \frac{\text{kg}}{\text{m}^3} = 5.6 \text{ kg}$$

11.

$$V = 3.0 \times 10^6 \text{ gal} \cdot \frac{3.785 \text{ L}}{1 \text{ gal}} \cdot \frac{1 \text{ m}^3}{1000 \text{ L}} = 1.14 \times 10^4 \text{ m}^3$$

$$\rho = 998 \ \frac{\text{kg}}{\text{m}^3}$$

$$F_w = ?$$

$$\rho = \frac{m}{V}$$

$$m = V\rho = 1.14 \times 10^4 \text{ m}^3 \cdot 998 \ \frac{\text{kg}}{\text{m}^3} = 1.14 \times 10^7 \text{ kg}$$

$$F_w = mg = 1.14 \times 10^7 \text{ kg} \cdot 9.80 \ \frac{\text{m}}{\text{s}^2} = 1.12 \times 10^8 \text{ N}$$

$$F_w = 1.12 \times 10^8 \text{ N} \cdot \frac{1 \text{ lb}}{4.45 \text{ N}} = 25{,}000{,}000 \text{ lb}$$

12.a.

Gold is more dense, so with cubes of equal size, the gold weighs more.

12.b.

For cubes of equal mass, the one with the lower density has to be larger—the lead.

13.

$$l = 50.0 \text{ m}$$

$$w = 25.0 \text{ m}$$

$$d = 2.00 \text{ m}$$

$$\rho = 998 \ \frac{\text{kg}}{\text{m}^3}$$

$$V = ?$$

$$F_w = ?$$

$$V = lwd = 50.0 \text{ m} \cdot 25.0 \text{ m} \cdot 2.00 \text{ m} = 2500.0 \text{ m}^3 \cdot \frac{1000 \text{ L}}{1 \text{ m}^3} \cdot \frac{1 \text{ gal}}{3.785 \text{ L}} = 6.61 \times 10^5 \text{ gal}$$

$$\rho = \frac{m}{V}$$

$$m = V\rho = 2500.0 \text{ m}^3 \cdot 998 \ \frac{\text{kg}}{\text{m}^3} = 2.495 \times 10^6 \text{ kg}$$

$$F_w = mg = 2.495 \times 10^6 \text{ kg} \cdot 9.80 \ \frac{\text{m}}{\text{s}^2} = 2.4451 \times 10^7 \text{ N} \cdot \frac{1 \text{ lb}}{4.45 \text{ N}} \cdot \frac{1 \text{ ton}}{2000 \text{ lb}} = 2750 \text{ tons}$$

Chapter 7

Temperature Unit Conversions

1.a.

$T_C = 32.0°C$

$T_F = ?$

$T_C = \dfrac{5}{9}(T_F - 32°)$

$T_F = \dfrac{9}{5}T_C + 32° = \dfrac{9}{5} \cdot 32.0°C + 32° = 89.6°F$

1.b.

$T_C = 32.0°C$

$T_K = ?$

$T_K = T_C + 273.2$

$T_K = 32.0°C + 273.2 = 305.2 \text{ K}$

2.a.

$T_F = 56.5°F$

$T_C = ?$

$T_C = \dfrac{5}{9}(T_F - 32°)$

$T_C = \dfrac{5}{9}(56.5°F - 32°) = 13.6°C$

2.b.

$T_F = 56.5°F$

$T_C = 13.6°C$

$T_K = ?$

$T_K = T_C + 273.2$

$T_K = 13.6°C + 273.2 = 286.8 \text{ K}$

3.a.

$T_K = 455.0 \text{ K}$

$T_C = 181.8°\text{C}$

$T_F = ?$

$T_C = \dfrac{5}{9}(T_F - 32°)$

$T_F = \dfrac{9}{5}T_C + 32°$

$T_F = \dfrac{9}{5} \cdot 181.8°\text{C} + 32° = 359.2°\text{F}$

3.b.

$T_K = 455.0 \text{ K}$

$T_C = ?$

$T_K = T_C + 273.2$

$T_C = T_K - 273.2 = 455.0 \text{ K} - 273.2 = 181.8°\text{C}$

4.a.

$T_F = -17.9°\text{F}$

$T_C = ?$

$T_C = \dfrac{5}{9}(T_F - 32°)$

$T_C = \dfrac{5}{9}(-17.9°\text{F} - 32°) = -27.7°\text{C}$

4.b.

$T_F = -17.9°\text{F}$

$T_C = -27.7°\text{C}$

$T_K = ?$

$T_K = T_C + 273.2$

$T_K = -27.7°\text{C} + 273.2 = 245.5 \text{ K}$

5.a.

$T_C = -41.6°C$

$T_F = ?$

$T_C = \dfrac{5}{9}(T_F - 32°)$

$T_F = \dfrac{9}{5}T_C + 32°$

$T_F = \left(\dfrac{9}{5} \cdot (-41.6°C)\right) + 32° = -42.9°F$

5.b.

$T_C = -41.6°C$

$T_K = ?$

$T_K = T_C + 273.2$

$T_K = -41.6°C + 273.2 = 231.6\ K$

6.a.

$T_K = 79.0\ K$

$T_C = ?$

$T_K = T_C + 273.2$

$T_C = T_K - 273.2$

$T_C = 79.0\ K - 273.2 = -194.2°C$

6.b.

$T_K = 79.0\ K$

$T_C = -194.2°C$

$T_F = ?$

$T_C = \dfrac{5}{9}(T_F - 32°)$

$T_F = \dfrac{9}{5}T_C + 32°$

$T_F = \left(\dfrac{9}{5} \cdot (-194.2°C)\right) + 32° = -317.6°\ F$

Chapter 8

Pressure Problems

1.

$$h = 33.5 \text{ ft} \cdot \frac{0.3048 \text{ m}}{1 \text{ ft}} = 10.2 \text{ m}$$

$$\rho = 998 \ \frac{\text{kg}}{\text{m}^3}$$

$$P = ?$$

$$P = \rho g h = 998 \ \frac{\text{kg}}{\text{m}^3} \cdot 9.80 \ \frac{\text{m}}{\text{s}^2} \cdot 10.2 \text{ m} = 99{,}760 \text{ Pa} \cdot \frac{1 \text{ kPa}}{1000 \text{ Pa}} = 99.8 \text{ kPa}$$

$$99{,}760 \text{ Pa} \cdot \frac{14.7 \text{ psi}}{101{,}325 \text{ Pa}} = 14.5 \text{ psi}$$

2.

$$h = 125 \text{ ft} \cdot \frac{0.3048 \text{ m}}{1 \text{ ft}} = 38.1 \text{ m}$$

$$\rho = 1.025 \ \frac{\text{g}}{\text{cm}^3} \cdot \frac{1 \text{ kg}}{1000 \text{ g}} \cdot \left(\frac{100 \text{ cm}}{1 \text{ m}} \right)^3 = 1{,}025 \ \frac{\text{kg}}{\text{m}^3}$$

$$P = ?$$

$$P = \rho g h = 1{,}025 \ \frac{\text{kg}}{\text{m}^3} \cdot 9.80 \ \frac{\text{m}}{\text{s}^2} \cdot 38.1 \text{ m} = 383{,}000 \text{ Pa} = 383 \text{ kPa}$$

$$382{,}715 \text{ Pa} \cdot \frac{14.7 \text{ psi}}{101{,}325 \text{ Pa}} = 55.5 \text{ psi}$$

3.

$$\rho = 998 \ \frac{\text{kg}}{\text{m}^3}$$

$$P = 101{,}325 \text{ Pa}$$

$$h = ?$$

$$P = \rho g h$$

$$h = \frac{P}{\rho g} = \frac{101{,}325 \text{ Pa}}{998 \ \frac{\text{kg}}{\text{m}^3} \cdot 9.80 \ \frac{\text{m}}{\text{s}^2}} = 10.4 \text{ m} \cdot \frac{1 \text{ ft}}{0.3048 \text{ m}} = 34.0 \text{ ft}$$

4.

$$h = 60.6 \text{ ft} \cdot \frac{0.3048 \text{ m}}{1 \text{ ft}} = 18.47 \text{ m}$$

$$\rho = 998 \, \frac{\text{kg}}{\text{m}^3}$$

$$P = ?$$

$$P = \rho g h = 998 \, \frac{\text{kg}}{\text{m}^3} \cdot 9.80 \, \frac{\text{m}}{\text{s}^2} \cdot 18.47 \text{ m} = 181{,}000 \text{ Pa} = 181 \text{ kPa}$$

5.

$$h = 18.47 \text{ m}$$

$$D = 10.0 \text{ ft} \cdot \frac{0.3048 \text{ m}}{1 \text{ ft}} = 3.048 \text{ m}$$

$$\rho = 998 \, \frac{\text{kg}}{\text{m}^3}$$

$$P = ?$$

$$V = \pi r^2 h = \pi \cdot \left(\frac{3.048 \text{ m}}{2} \right)^2 \cdot 18.47 \text{ m} = 134.8 \text{ m}^3$$

$$\rho = \frac{m}{V}$$

$$m = \rho V = 998 \, \frac{\text{kg}}{\text{m}^3} \cdot 134.8 \text{ m}^3 = 134{,}500 \text{ kg}$$

$$F_w = mg = 134{,}500 \text{ kg} \cdot 9.80 \, \frac{\text{m}}{\text{s}^2} = 1{,}318{,}000 \text{ N}$$

$$A = \pi r^2 = \pi \left(\frac{3.048 \text{ m}}{2} \right)^2 = 7.297 \text{ m}^2$$

$$P = \frac{F_w}{A} = \frac{1{,}318{,}000 \text{ N}}{7.297 \text{ m}^2} = 181{,}000 \text{ Pa} = 181 \text{ kPa}$$

6.

$$P = 181{,}000 \text{ Pa}$$

$$P_{atm(atmospheric)} = 101{,}325 \text{ Pa}$$

$$P_{abs(absolute)} = ?$$

$$P_{abs} = P_{gauge} + P_{atm} = 181{,}000 \text{ Pa} + 101{,}325 \text{ Pa} = 282 \text{ kPa}$$

$$282{,}000 \text{ Pa} \cdot \frac{14.7 \text{ psi}}{101{,}325 \text{ Pa}} = 40.9 \text{ psi}$$

7.

$$P_{gauge} = 55.0 \text{ psi}$$

$$P_{abs_i} = 40.9 \text{ psi}$$

$$P_{abs_f} = ?$$

$$P_{abs_f} = P_{abs_i} + P_{gauge} = 40.9 \text{ psi} + 55.0 \text{ psi} = 95.9 \text{ psi}$$

$$95.9 \text{ psi} \cdot \frac{101,325 \text{ Pa}}{14.7 \text{ psi}} = 661,000 \text{ Pa} = 661 \text{ kPa}$$

8.a.

$$P_{gauge} = 14.5 \text{ psi}$$

$$P_{atm} = 14.7 \text{ psi}$$

$$P_{abs} = P_{gauge} + P_{atm} = 14.5 \text{ psi} + 14.7 \text{ psi} = 29.2 \text{ psi}$$

$$29.2 \text{ psi} \cdot \frac{101,325 \text{ Pa}}{14.7 \text{ psi}} = 201,200 \text{ Pa} = 201.2 \text{ kPa}$$

8.b.

$$P_{gauge} = 14.5 \text{ psi}$$

$$P_{atm} = 14.7 \text{ psi}$$

$$P_{abs} = P_{gauge} + P_{atm} = 55.5 \text{ psi} + 14.7 \text{ psi} = 70.2 \text{ psi}$$

$$70.2 \text{ psi} \cdot \frac{101,325 \text{ Pa}}{14.7 \text{ psi}} = 484,000 \text{ Pa} = 484 \text{ kPa}$$

9.

$$P = 83.4 \text{ kPa} \cdot \frac{1000 \text{ Pa}}{1 \text{ kPa}} = 83,400 \text{ Pa}$$

$$\rho = 13,600 \ \frac{\text{kg}}{\text{m}^3}$$

$$h = ?$$

$$P = \rho g h$$

$$h = \frac{P}{\rho g} = \frac{83,400 \text{ Pa}}{13,600 \ \frac{\text{kg}}{\text{m}^3} \cdot 9.80 \ \frac{\text{m}}{\text{s}^2}} = 0.626 \text{ m} = 626 \text{ mm}$$

10.

$$P = 2{,}800 \text{ psi} \cdot \frac{101{,}325 \text{ Pa}}{14.7 \text{ psi}} = 1.93 \times 10^7 \text{ Pa}$$

$$F_w = 2{,}200 \text{ lb} \cdot \frac{4.45 \text{ N}}{1 \text{ lb}} = 9{,}790 \text{ N}$$

$$D = 0.75 \text{ in} \cdot \frac{2.54 \text{ cm}}{1 \text{ in}} \cdot \frac{1 \text{ m}}{100 \text{ cm}} = 0.01905 \text{ m}$$

$$P_{foot} = ?$$

P_{foot} is the pressure under each foot.

$$r = \frac{D}{2} = \frac{0.01905 \text{ m}}{2} = 0.009525 \text{ m}$$

$$A = \pi r^2 = \pi (0.009525 \text{ m})^2 = 2.850 \times 10^{-4} \text{ m}^2$$

Since there are four feet, the total pressure will be divided by four.

$$P_{foot} = \frac{F}{4A} = \frac{9{,}790 \text{ N}}{4 \cdot 2.850 \times 10^{-4} \text{ m}^2} = 8{,}588{,}000 \text{ Pa} \cdot \frac{14.7 \text{ psi}}{101{,}325 \text{ Pa}} = 1{,}200 \text{ psi}$$

11.

$$P = 3{,}200 \text{ psi} \cdot 0.75 = 2{,}400 \text{ psi} \cdot \frac{101{,}325 \text{ Pa}}{14.7 \text{ psi}} = 1.65 \times 10^7 \text{ Pa}$$

$$\rho = 998 \ \frac{\text{kg}}{\text{m}^3}$$

$$h = ?$$

$$P = \rho g h$$

$$h = \frac{P}{\rho g} = \frac{1.65 \times 10^7 \text{ Pa}}{998 \ \frac{\text{kg}}{\text{m}^3} \cdot 9.80 \ \frac{\text{m}}{\text{s}^2}} = 1{,}700 \text{ m}$$

12.

$$A = 120 \text{ cm}^2 \cdot \left(\frac{1 \text{ m}}{100 \text{ cm}}\right)^2 = 0.012 \text{ m}^2$$

$$F_w = 195 \text{ lb} \cdot \frac{4.45 \text{ N}}{1 \text{ lb}} = 867.8 \text{ N}$$

$$P_{shoes} = ?$$

$$P_{shoes} = \frac{F}{2A} = \frac{867.8 \text{ N}}{2 \cdot 0.012 \text{ m}^2} = 36,000 \text{ Pa}$$

$$36,000 \text{ Pa} \cdot \frac{14.7 \text{ psi}}{101,325 \text{ Pa}} = 5.2 \text{ psi}$$

13.

$$l = 0.750 \text{ cm} \cdot \frac{1 \text{ m}}{100 \text{ cm}} = 0.00750 \text{ m}$$

$$w = 0.750 \text{ cm} \cdot \frac{1 \text{ m}}{100 \text{ cm}} = 0.00750 \text{ m}$$

$$F_w = 130 \text{ lb} \cdot \frac{4.45 \text{ N}}{1 \text{ lb}} = 579 \text{ N}$$

$$P_{heels} = ?$$

$$P_{heels} = \frac{F_w}{2A}$$

$$A = lw = (0.00750 \text{ m})^2 = 5.60 \times 10^{-5} \text{ m}^2$$

$$P_{heels} = \frac{579 \text{ N}}{2 \cdot 5.60 \times 10^{-5} \text{ m}^2} = 5,200,000 \text{ Pa}$$

$$5,200,000 \text{ Pa} \cdot \frac{14.7 \text{ psi}}{101,325 \text{ Pa}} = 750 \text{ psi}$$

14.

$$h = 20.0 \text{ ft} \cdot \frac{0.3048 \text{ m}}{1 \text{ ft}} = 6.096 \text{ m}$$

$$\rho = 1.025 \frac{\text{g}}{\text{cm}^3} \cdot \frac{1 \text{ kg}}{1000 \text{ g}} \cdot \frac{1 \times 10^6 \text{ cm}^3}{\text{m}^3} = 1025 \frac{\text{kg}}{\text{m}^3}$$

$$P_{abs} = ?$$

$$P = \rho g h = 1025 \frac{\text{kg}}{\text{m}^3} \cdot 9.80 \frac{\text{m}}{\text{s}^2} \cdot 6.096 \text{ m} = 6.123 \times 10^4 \text{ Pa}$$

$$P_{abs} = P_{gauge} + P_{atm} = 6.123 \times 10^4 \text{ Pa} + 101,325 \text{ Pa} = 1.626 \times 10^5 \text{ Pa}$$

$$P_{abs} = 1.626 \times 10^5 \text{ Pa} \cdot \frac{14.7 \text{ psi}}{101,325 \text{ Pa}} = 23.6 \text{ psi}$$

Buoyancy Problems

1.

$$w = 1.5 \text{ in} \cdot \frac{2.54 \text{ cm}}{1 \text{ in}} \cdot \frac{1 \text{ m}}{100 \text{ cm}} = 0.0381 \text{ m}$$

$$h = 3.5 \text{ in} \cdot \frac{2.54 \text{ cm}}{1 \text{ in}} \cdot \frac{1 \text{ m}}{100 \text{ cm}} = 0.0889 \text{ m}$$

$$l = 12 \text{ in} \cdot \frac{2.54 \text{ cm}}{1 \text{ in}} \cdot \frac{1 \text{ m}}{100 \text{ cm}} = 0.3048 \text{ m}$$

$$\rho = 998 \frac{\text{kg}}{\text{m}^3}$$

$$F_{B(Buoyant)} = ?$$

$$V = lwh = 0.0381 \text{ m} \cdot 0.0889 \text{ m} \cdot 0.3048 \text{ m} = 1.03 \times 10^{-3} \text{ m}^3$$

$$\rho = \frac{m}{V}$$

$$m = V\rho = 1.03 \times 10^{-3} \text{ m}^3 \cdot 998 \frac{\text{kg}}{\text{m}^3} = 1.03 \text{ kg}$$

$$F_w = mg = 1.03 \text{ kg} \cdot 9.80 \frac{\text{m}}{\text{s}^2} = 1.0 \times 10^1 \text{ N}$$

$$10.1 \text{ N} \cdot \frac{1 \text{ lb}}{4.45 \text{ N}} = 2.3 \text{ lb}$$

2.

$V = 1.03 \times 10^{-3} \text{ m}^3$

$\rho_{saltwater} = 1.025 \, \dfrac{\text{g}}{\text{cm}^3} \cdot \dfrac{1 \text{ kg}}{1000 \text{ g}} \cdot \left(\dfrac{100 \text{ cm}}{1 \text{ m}} \right)^3 = 1{,}025 \, \dfrac{\text{kg}}{\text{m}^3}$

$\rho_{mercury} = 13.6 \, \dfrac{\text{g}}{\text{cm}^3} \cdot \dfrac{1 \text{ kg}}{1000 \text{ g}} \cdot \left(\dfrac{100 \text{ cm}}{1 \text{ m}} \right)^3 = 1.36 \times 10^4 \, \dfrac{\text{kg}}{\text{m}^3}$

$F_{B,\,saltwater} = ?$

$F_{B,\,mercury} = ?$

$\rho_{saltwater} = \dfrac{m}{V}$

$m = \rho_{saltwater} V = 1{,}025 \, \dfrac{\text{kg}}{\text{m}^3} \cdot 1.03 \times 10^{-3} \text{ m}^3 = 1.056 \text{ kg}$

$F_{B,\,saltwater} = F_{w,\,saltwater} = mg = 1.056 \text{ kg} \cdot 9.80 \, \dfrac{\text{m}}{\text{s}^2} = 1.0 \times 10^1 \text{ N}$

$F_{B,\,saltwater} = 10.3 \text{ N} \cdot \dfrac{1 \text{ lb}}{4.45 \text{ N}} = 2.3 \text{ lb}$

$\rho_{mercury} = \dfrac{m}{V}$

$m = \rho_{mercury} V = 1.36 \times 10^4 \, \dfrac{\text{kg}}{\text{m}^3} \cdot 1.03 \times 10^{-3} \text{ m}^3 = 14.0 \text{ kg}$

$F_{B,\,mercury} = F_{w,\,mercury} = mg = 14.0 \text{ kg} \cdot 9.80 \, \dfrac{\text{m}}{\text{s}^2} = 140 \text{ N}$

$F_{B,\,mercury} = 140 \text{ N} \cdot \dfrac{1 \text{ lb}}{4.45 \text{ N}} = 31 \text{ lb}$

3.

$$D = 22.5 \text{ in} \cdot \frac{2.54 \text{ cm}}{1 \text{ in}} \cdot \frac{1 \text{ m}}{100 \text{ cm}} = 0.5715 \text{ m}$$

$$h = 33.5 \text{ in} \cdot \frac{2.54 \text{ cm}}{1 \text{ in}} \cdot \frac{1 \text{ m}}{100 \text{ cm}} = 0.8509 \text{ m}$$

$$F_w = 54.0 \text{ lb} \cdot \frac{4.45 \text{ N}}{1 \text{ lb}} = 240.3 \text{ N}$$

$$F_{cable} = ?$$

$$V = \pi \left(\frac{D}{2} \right)^2 h = \pi \left(\frac{0.5715 \text{ m}}{2} \right)^2 \cdot 0.8509 \text{ m} = 0.2187 \text{ m}^3$$

$$\rho = \frac{m}{V}$$

$$m = \rho V = 998 \ \frac{\text{kg}}{\text{m}^3} \cdot 0.2187 \text{ m}^3 = 218.3 \text{ kg}$$

$$F_w = mg = 218.3 \text{ kg} \cdot 9.80 \ \frac{\text{m}}{\text{s}^2} = 2.139 \times 10^3 \text{ N}$$

$$F_B = F_w + F_a$$

$$F_a = F_{cable} = F_B - F_w = 2.139 \times 10^3 \text{ N} - 240.3 \text{ N} = 1.90 \times 10^3 \text{ N}$$

4.

$$\rho_{s(styrofoam)} = 55 \ \frac{\text{kg}}{\text{m}^3}$$

$$h = 12 \ \text{in} \cdot \frac{2.54 \ \text{cm}}{1 \ \text{in}} \cdot \frac{1 \ \text{m}}{100 \ \text{cm}} = 0.305 \ \text{m}$$

$$l = 8.0 \ \text{ft} \cdot \frac{0.3048 \ \text{m}}{1 \ \text{ft}} = 2.44 \ \text{m}$$

$$w = 6.0 \ \text{ft} \cdot \frac{0.3048 \ \text{m}}{1 \ \text{ft}} = 1.83 \ \text{m}$$

$$F_{a(additional)} = ?$$

$$V_s = 0.305 \ \text{m} \cdot 2.44 \ \text{m} \cdot 1.83 \ \text{m} = 1.36 \ \text{m}^3$$

$$\rho_s = \frac{m_s}{V_s}$$

$$m_s = \rho_s V_s = 55 \ \frac{\text{kg}}{\text{m}^3} \cdot 1.36 \ \text{m}^3 = 74.8 \ \text{kg}$$

$$F_w = m_s g = 74.8 \ \text{kg} \cdot 9.80 \ \frac{\text{m}}{\text{s}^2} = 733 \ \text{N}$$

$$V_{dw(displaced \ water)} = \frac{V_{block}}{2} = 0.681 \ \text{m}^3$$

$$m_{dw} = \rho_{water} V_{dw} = 998 \ \frac{\text{kg} \cdot \text{m}}{\text{s}} \cdot 0.681 \ \text{m}^3 = 679.6 \ \text{kg}$$

$$F_B = m_{dw} g = 679.6 \ \text{kg} \cdot 9.80 \ \frac{\text{m}}{\text{s}^2} = 6.66 \times 10^3 \ \text{N}$$

$$F_B = F_w + F_a$$

$$F_a = F_B - F_w = 6.66 \times 10^3 \ \text{N} - 733 \ \text{N} = 5,900 \ \text{N}$$

$$5,927 \ \text{N} \cdot \frac{1 \ \text{lb}}{4.45 \ \text{N}} = 1,300 \ \text{lb}$$

5.

$$F_{w,\,dock} = 450 \text{ lb} \cdot \frac{4.45 \text{ N}}{1 \text{ lb}} = 2,002 \text{ N}$$

$$V_{drums} = 8 \cdot 0.219 \text{ m}^3 = 1.75 \text{ m}^3$$

$$F_{w,\,drums} = 8 \cdot 54.0 \text{ lb} \cdot \frac{4.45 \text{ N}}{1 \text{ lb}} = 1,922 \text{ N}$$

$$F_a = ?$$

$$F_{w,\,total} = F_{w,\,drums} + F_{w,\,dock} = 2,002 \text{ N} + 1,922 \text{ N} = 3,924 \text{ N}$$

$$V_{dw(displaced\ water)} = \frac{V_{drums}}{2} = \frac{1.75 \text{ m}^3}{2} = 0.875 \text{ m}^3$$

$$\rho_{water} = \frac{m_{dw}}{V_{dw}}$$

$$m_{dw} = V_{dw}\rho_{water} = 0.875 \text{ m}^3 \cdot 998 \ \frac{\text{kg}}{\text{m}^3} = 873 \text{ kg}$$

$$F_B = m_{dw}g = 873 \text{ kg} \cdot 9.80 \ \frac{\text{m}}{\text{s}^2} = 8,555 \text{ N}$$

$$F_B = F_w + F_a$$
$$F_a = F_B - F_w = 8,555 \text{ N} - 3,924 \text{ N} = 4,600 \text{ N}$$

$$4,631 \text{ N} \cdot \frac{1 \text{ lb}}{4.45 \text{ N}} = 1.0 \times 10^3 \text{ lb}$$

Chapter 9

1.

$f = 60.0 \text{ Hz}$

$\tau = ?$

$\tau = \dfrac{1}{f}$

$\tau = \dfrac{1}{60.0 \text{ Hz}} = 0.0167 \text{ s}$

2.

$\tau = 2.155 \times 10^{-5} \text{ s}$

$f = ?$

$\tau = \dfrac{1}{f}$

$f = \dfrac{1}{\tau} = \dfrac{1}{2.155 \times 10^{-5} \text{ s}} = 46,404 \text{ Hz} \cdot \dfrac{1 \text{ kHz}}{1000 \text{ Hz}} = 46.40 \text{ kHz}$

3.

$f = 26.0 \text{ kHz} \cdot \dfrac{1000 \text{ Hz}}{1 \text{ kHz}} = 2.60 \times 10^{4} \text{ Hz}$

$\tau = ?$

$\tau = \dfrac{1}{f} = \dfrac{1}{2.60 \times 10^{4} \text{ Hz}} = 3.85 \times 10^{-5} \text{ s}$

4.

$f = 2.60 \times 10^{4} \text{ Hz}$

$v = 342 \; \dfrac{\text{m}}{\text{s}}$

$\lambda = ?$

$v = \lambda f$

$\lambda = \dfrac{v}{f} = \dfrac{342 \; \dfrac{\text{m}}{\text{s}}}{2.60 \times 10^{4} \text{ Hz}} = 1.32 \times 10^{-2} \text{ m} \cdot \dfrac{100 \text{ cm}}{1 \text{ m}} = 1.32 \text{ cm}$

5.

$$f = 89.5 \text{ MHz} \cdot \frac{1 \times 10^6 \text{ Hz}}{1 \text{ MHz}} = 8.95 \times 10^7 \text{ Hz}$$

$$v = c = 3.00 \times 10^8 \ \frac{\text{m}}{\text{s}}$$

$$\lambda = ?$$

$$\tau = ?$$

$$v = \lambda f$$

$$\lambda = \frac{v}{f} = \frac{3.00 \times 10^8 \ \frac{\text{m}}{\text{s}}}{8.95 \times 10^7 \text{ Hz}} = 3.35 \text{ m}$$

$$\tau = \frac{1}{f} = \frac{1}{8.95 \times 10^7 \text{ Hz}} = 1.12 \times 10^{-8} \text{ s} \cdot \frac{1 \times 10^6 \ \mu\text{s}}{1 \text{ s}} = 0.0112 \ \mu\text{s}$$

6.

$$v = c = 3.00 \times 10^8 \ \frac{\text{m}}{\text{s}}$$

$$f = 1310 \text{ kHz} \cdot \frac{1000 \text{ Hz}}{1 \text{ kHz}} = 1.31 \times 10^6 \text{ Hz}$$

$$\lambda = ?$$

$$\tau = ?$$

$$v = \lambda f$$

$$\lambda = \frac{v}{f} = \frac{3.00 \times 10^8 \ \frac{\text{m}}{\text{s}}}{1.31 \times 10^6 \text{ Hz}} = 229 \text{ m}$$

$$\tau = \frac{1}{f} = \frac{1}{1.31 \times 10^6 \text{ Hz}} = 7.64 \times 10^{-7} \text{ s} \cdot \frac{1 \times 10^6 \ \mu\text{s}}{1 \text{ s}} = 0.763 \ \mu\text{s}$$

7.

$$v = c = 3.00 \times 10^8 \ \frac{m}{s}$$

$$\lambda = 542 \ nm \cdot \frac{1 \ m}{1 \times 10^9 \ nm} = 5.42 \times 10^{-7} \ m$$

$$\tau = ?$$

$$f = ?$$

$$v = \lambda f$$

$$f = \frac{v}{\lambda} = \frac{3.00 \times 10^8 \ \frac{m}{s}}{5.42 \times 10^{-7} \ m} = 5.54 \times 10^{14} \ Hz \cdot \frac{1 \ THz}{1 \times 10^{12} \ Hz} = 554 \ THz$$

$$\tau = \frac{1}{f} = \frac{1}{5.54 \times 10^{14} \ Hz} = 1.81 \times 10^{-15} \ s \cdot \frac{1 \times 10^{12} \ ps}{1 \ s} = 0.00181 \ ps$$

8.

$$\lambda = 10.6 \ \mu m \cdot \frac{1 \ m}{1 \times 10^6 \ \mu m} = 1.06 \times 10^{-5} \ m$$

$$v = 3.00 \times 10^8 \ \frac{m}{s}$$

$$\tau = ?$$

$$f = ?$$

$$v = \lambda f$$

$$f = \frac{v}{\lambda} = \frac{3.00 \times 10^8 \ \frac{m}{s}}{1.06 \times 10^{-5} \ m} = 2.83 \times 10^{13} \ Hz \cdot \frac{1 \ GHz}{1 \times 10^9 \ Hz} = 2.83 \times 10^4 \ GHz$$

$$\tau = \frac{1}{f} = \frac{1}{2.83 \times 10^{13} \ Hz} = 3.53 \times 10^{-14} \ s \cdot \frac{1 \times 10^9 \ ns}{1 \ s} = 3.53 \times 10^{-5} \ ns$$

9.

$$f = 33 \text{ kHz} \cdot \frac{1000 \text{ Hz}}{1 \text{ kHz}} = 3.3 \times 10^4 \text{ Hz}$$

$$v = 342 \; \frac{\text{m}}{\text{s}}$$

$$\tau = ?$$

$$\lambda = ?$$

$$\tau = \frac{1}{f} = \frac{1}{3.3 \times 10^4 \text{ Hz}} = 3.03 \times 10^{-5} \text{ s} \cdot \frac{1000 \text{ ms}}{1 \text{ s}} = 0.030 \text{ ms}$$

$$v = \lambda f$$

$$\lambda = \frac{v}{f} = \frac{342 \; \frac{\text{m}}{\text{s}}}{3.3 \times 10^4 \text{ Hz}} = 1.04 \times 10^{-2} \text{ m} \cdot \frac{1000 \text{ mm}}{1 \text{ m}} = 1.0 \times 10^1 \text{ mm}$$

10.

Wavelengths smaller than the hole size pass through the screen.

$$v = 3.00 \times 10^8 \; \frac{\text{m}}{\text{s}}$$

$$\lambda = 2.00 \text{ mm} \cdot \frac{1 \text{ m}}{1000 \text{ mm}} = 2.00 \times 10^{-3} \text{ m}$$

$$v = \lambda f$$

$$f = \frac{v}{\lambda} = \frac{3.00 \times 10^8 \; \frac{\text{m}}{\text{s}}}{2.00 \times 10^{-3} \text{ m}} = 1.50 \times 10^{11} \text{ Hz}$$

Thus, when $f > 1.50 \times 10^{11}$ Hz, the waves pass through.

11.

$$\lambda_b = 470 \text{ nm} \cdot \frac{1 \text{ m}}{1 \times 10^9 \text{ nm}} = 4.7 \times 10^{-7} \text{ m}$$

$$\lambda_g = 550 \text{ nm} \cdot \frac{1 \text{ m}}{1 \times 10^9 \text{ nm}} = 5.5 \times 10^{-7} \text{ m}$$

$$\lambda_r = 680 \text{ nm} \cdot \frac{1 \text{ m}}{1 \times 10^9 \text{ nm}} = 6.8 \times 10^{-7} \text{ m}$$

$$v = 3.00 \times 10^8 \ \frac{\text{m}}{\text{s}}$$

$$f_b = ?$$

$$f_g = ?$$

$$f_r = ?$$

$$v = \lambda f$$

$$f = \frac{v}{\lambda}$$

$$f_b = \frac{3.00 \times 10^8 \ \frac{\text{m}}{\text{s}}}{4.7 \times 10^{-7} \text{ m}} = 6.4 \times 10^{14} \text{ Hz}$$

$$f_g = \frac{3.00 \times 10^8 \ \frac{\text{m}}{\text{s}}}{5.5 \times 10^{-7} \text{ m}} = 5.5 \times 10^{14} \text{ Hz}$$

$$f_r = \frac{3.00 \times 10^8 \ \frac{\text{m}}{\text{s}}}{6.8 \times 10^{-7} \text{ m}} = 4.4 \times 10^{14} \text{ Hz}$$

12.

$$f = 20.00 \text{ Hz}$$

$$v = 342.0 \ \frac{\text{m}}{\text{s}}$$

$$\lambda = ?$$

$$v = \lambda f$$

$$\lambda = \frac{v}{f} = \frac{342.0 \ \frac{\text{m}}{\text{s}}}{20.00 \text{ Hz}} = 17.10 \text{ m}$$

13.a.

$$f = 4.67 \times 10^{20} \text{ Hz}$$

$$v = 3.00 \times 10^8 \ \frac{\text{m}}{\text{s}}$$

$$\lambda = ?$$

$$v = \lambda f$$

$$\lambda = \frac{v}{f} = \frac{3.00 \times 10^8 \ \frac{\text{m}}{\text{s}}}{4.67 \times 10^{20} \text{ Hz}} = 6.42 \times 10^{-13} \text{ m} \cdot \frac{1 \times 10^{12} \text{ pm}}{\text{m}} = 0.642 \text{ pm}$$

13.b.

$$f = 9.9876 \times 10^{18} \text{ Hz}$$

$$v = 3.00 \times 10^8 \ \frac{\text{m}}{\text{s}}$$

$$\lambda = ?$$

$$v = \lambda f$$

$$\lambda = \frac{v}{f} = \frac{3.00 \times 10^8 \ \frac{\text{m}}{\text{s}}}{9.9876 \times 10^{18} \text{ Hz}} = 3.00 \times 10^{-11} \text{ m} \cdot \frac{1 \times 10^{9} \text{ nm}}{\text{m}} = 0.0300 \text{ nm}$$

13.c.

$$f = 2.555 \times 10^{10} \text{ Hz}$$

$$v = 3.00 \times 10^8 \ \frac{\text{m}}{\text{s}}$$

$$\lambda = ?$$

$$v = \lambda f$$

$$\lambda = \frac{v}{f} = \frac{3.00 \times 10^8 \ \frac{\text{m}}{\text{s}}}{2.555 \times 10^{10} \text{ Hz}} = 0.0117 \text{ m} \cdot \frac{1000 \text{ mm}}{\text{m}} = 11.7 \text{ mm}$$

13.d.

$f = 1.172 \times 10^{15} \text{ Hz}$

$v = 3.00 \times 10^8 \ \dfrac{\text{m}}{\text{s}}$

$\lambda = ?$

$v = \lambda f$

$\lambda = \dfrac{v}{f} = \dfrac{3.00 \times 10^8 \ \dfrac{\text{m}}{\text{s}}}{1.172 \times 10^{15} \text{ Hz}} = 2.56 \times 10^{-7} \text{ m} \cdot \dfrac{1 \times 10^9 \text{ nm}}{\text{m}} = 256 \text{ nm}$

13.e.

$f = 2.83 \times 10^{13} \text{ Hz}$

$v = 3.00 \times 10^8 \ \dfrac{\text{m}}{\text{s}}$

$\lambda = ?$

$v = \lambda f$

$\lambda = \dfrac{v}{f} = \dfrac{3.00 \times 10^8 \ \dfrac{\text{m}}{\text{s}}}{2.83 \times 10^{13} \text{ Hz}} = 1.06 \times 10^{-5} \text{ m} \cdot \dfrac{1 \times 10^6 \ \mu\text{m}}{\text{m}} = 10.6 \ \mu\text{m}$

14.

$f = 2.45 \text{ GHz} \cdot \dfrac{1 \times 10^9 \text{ Hz}}{1 \text{ GHz}} = 2.45 \times 10^9 \text{ Hz}$

$D = 3 \text{ mm} \cdot \dfrac{1 \text{ m}}{10^3 \text{ mm}} = 0.003 \text{ m}$

$v = \lambda f$

$\lambda = \dfrac{v}{f} = \dfrac{3.00 \times 10^8 \ \dfrac{\text{m}}{\text{s}}}{2.45 \times 10^9 \text{ Hz}} = 0.12 \text{ m}$

$ratio = ?$

$ratio = \dfrac{\lambda}{D} = \dfrac{0.12 \text{ m}}{0.003 \text{ m}} = 40$

15.a.

$$f = 1.00 \text{ kHz} \cdot \frac{1000 \text{ Hz}}{1 \text{ kHz}} = 1.00 \times 10^3 \text{ Hz}$$

$$v = 342 \; \frac{\text{m}}{\text{s}}$$

$$\lambda = ?$$

$$v = \lambda f$$

$$\lambda = \frac{v}{f} = \frac{342 \; \frac{\text{m}}{\text{s}}}{1.00 \times 10^3 \text{ Hz}} = 0.342 \text{ m}$$

15.b.

$$f = 1.00 \text{ kHz} \cdot \frac{1000 \text{ Hz}}{1 \text{ kHz}} = 1.00 \times 10^3 \text{ Hz}$$

$$v = 1402 \; \frac{\text{m}}{\text{s}}$$

$$\lambda = ?$$

$$v = \lambda f$$

$$\lambda = \frac{v}{f} = \frac{1402 \; \frac{\text{m}}{\text{s}}}{1.00 \times 10^3 \text{ Hz}} = 1.40 \text{ m}$$

15.c.

$$f = 1.00 \text{ kHz} \cdot \frac{1000 \text{ Hz}}{1 \text{ kHz}} = 1.00 \times 10^3 \text{ Hz}$$

$$v = 5130 \; \frac{\text{m}}{\text{s}}$$

$$\lambda = ?$$

$$v = \lambda f$$

$$\lambda = \frac{v}{f} = \frac{5130 \; \frac{\text{m}}{\text{s}}}{1.00 \times 10^3 \text{ Hz}} = 5.13 \text{ m}$$

15.d.

$$f = 1.00 \text{ kHz} \cdot \frac{1000 \text{ Hz}}{1 \text{ kHz}} = 1.00 \times 10^3 \text{ Hz}$$

$$v = 965 \frac{\text{m}}{\text{s}}$$

$$\lambda = ?$$

$$v = \lambda f$$

$$\lambda = \frac{v}{f} = \frac{965 \frac{\text{m}}{\text{s}}}{1.00 \times 10^3 \text{ Hz}} = 0.965 \text{ m}$$

Chapter 11

Introductory Circuit Calculations

1.

$I = 13.00$ A

$V = 25.00$ V

$R = ?$

$V = IR$

$R = \dfrac{V}{I} = \dfrac{25.00 \text{ V}}{13.00 \text{ A}} = 1.923 \ \Omega$

2.

$V = 24$ V

$R = 250 \ \Omega$

$I = ?$

$V = IR$

$I = \dfrac{V}{R} = \dfrac{24 \text{ V}}{250 \ \Omega} = 0.096 \text{ A} \cdot \dfrac{1000 \text{ mA}}{1 \text{ A}} = 96 \text{ mA}$

3.

$R = 12.20$ kΩ

$V = 4.500$ V

$I = ?$

$V = IR$

$I = \dfrac{V}{R} = \dfrac{4.500 \text{ V}}{12.20 \text{ k}\Omega} = 0.3689 \text{ mA}$

4.

$I = 0.0300$ mA

$R = 33.3 \text{ M}\Omega \cdot \dfrac{1 \times 10^3 \text{ k}\Omega}{1 \text{ M}\Omega} = 3.33 \times 10^4 \text{ k}\Omega$

$V = ?$

$V = IR = 0.0300 \text{ mA} \cdot 3.33 \times 10^4 \text{ k}\Omega = 999 \text{ V}$

5.a.

$I = 13.00$ A

$V = 25.00$ V

$P = ?$

$P = VI = 25.00 \text{ V} \cdot 13.00 \text{ A} = 325.0$ W

5.b.

$V = 24$ V

$I = 96$ mA

$P = ?$

$P = VI = 24 \text{ V} \cdot 96 \text{ mA} = 2300 \text{ mW} = 2.3$ W

5.c.

$V = 4.500$ V

$I = 0.3689$ mA

$P = ?$

$P = VI = 4.500 \text{ V} \cdot 0.3689 \text{ mA} = 1.660$ mW

6.

$V = 120$ V

$P = 60.00$ W

$R = ?$

$I = ?$

$P = VI$

$I = \dfrac{P}{V} = \dfrac{60.00 \text{ W}}{120 \text{ V}} = 0.50$ A

$V = IR$

$R = \dfrac{V}{I} = \dfrac{120 \text{ V}}{0.50 \text{ A}} = 240 \ \Omega$

7.

$V = 120$ V

$I = 12$ A

$P = ?$

$P = VI = 120 \text{ V} \cdot 12 \text{ A} = 1440 \text{ W} \cdot \dfrac{1 \text{ kW}}{1000 \text{ W}} = 1.4$ kW

8.

$I = 13.5\ \mu\text{A}$

$V = 6.0\ \text{V}$

$P = ?$

$P = VI = 6.0\ \text{V} \cdot 13.5\ \mu\text{A} = 81\ \mu\text{W}$

9.

$$P = 155\ \text{MW} \cdot \frac{1 \times 10^6\ \text{W}}{1\ \text{MW}} = 1.55 \times 10^8\ \text{W}$$

$V = 762\ \text{V}$

$I = ?$

$P = VI$

$$I = \frac{P}{V} = \frac{1.55 \times 10^8\ \text{W}}{762\ \text{V}} = 203{,}000\ \text{A}$$

Equivalent Resistance Exercises

1.

series

$980\ \Omega + 440\ \Omega = 1420\ \Omega$

2.

series

$13\ \text{k}\Omega + 22\ \text{k}\Omega = 35\ \text{k}\Omega$

3.

parallel

$$\frac{5.1 \text{ M}\Omega \cdot 0.99 \text{ M}\Omega}{5.1 \text{ M}\Omega + 0.99 \text{ M}\Omega} = 0.8291 \text{ M}\Omega$$

4.

parallel

$$\frac{13.5 \text{ k}\Omega \cdot 6.2 \text{ k}\Omega}{13.5 \text{ k}\Omega + 6.2 \text{ k}\Omega} = 4.2487 \text{ k}\Omega$$

5.

parallel

$$\frac{12 \text{ k}\Omega \cdot 12 \text{ k}\Omega}{12 \text{ k}\Omega + 12 \text{ k}\Omega} = 6 \text{ k}\Omega$$

6.

R_1 and R_2 are in series.

R_3 and R_4 are in series.

R_6 and R_7 are in parallel.

R_8 and R_9 are in parallel.

Equivalent Resistance Calculations

1.

$$R_3 + R_4 = 47\ \Omega + 130\ \Omega = 177\ \Omega$$

$$R_2 \parallel 177\ \Omega = \frac{32\ \Omega \cdot 177\ \Omega}{32\ \Omega + 177\ \Omega} = 27.1005\ \Omega$$

$$R_{EQ} = R_1 + 27.1005\ \Omega_4 = 55\ \Omega + 27.1005\ \Omega = 82.1005\ \Omega$$

2.

$$R_3 \parallel R_4 = \frac{2.2\ \text{k} \cdot 0.990\ \text{k}}{2.2\ \text{k} + 0.990\ \text{k}} = 0.6828\ \text{k}$$

$$R_2 \parallel 0.6828\ \text{k} = \frac{5.7\ \text{k} \cdot 0.6828\ \text{k}}{5.7\ \text{k} + 0.6828\ \text{k}} = 0.6098\ \text{k}$$

$$R_{EQ} = R_1 \parallel 0.6098\ \text{k} = \frac{4.1\ \text{k} \cdot 0.6098\ \text{k}}{4.1\ \text{k} + 0.6098\ \text{k}} = 0.5308\ \text{k}\Omega$$

3.

$$R_3 \parallel R_4 = \frac{33\ \text{k} \cdot 33\ \text{k}}{33\ \text{k} + 33\ \text{k}} = 16.5\ \text{k}$$

$$R_{EQ} = R_1 + R_2 + 16.5\ \text{k} = 13\ \text{k} + 27\ \text{k} + 16.5\ \text{k} = 56.5000\ \text{k}\Omega$$

4.

$$R_4 + R_5 = 630 \ \Omega + 630 \ \Omega = 1260 \ \Omega$$

$$R_3 \| 1260 \ \text{k} = \frac{470 \ \Omega \cdot 1260 \ \Omega}{470 \ \Omega + 1260 \ \Omega} = 342.3121 \ \Omega$$

$$R_2 + 342.3121 \ \Omega = 220 \ \Omega + 342.3121 \ \Omega = 562.3121 \ \Omega$$

$$R_{EQ} = R_1 \| 562.3121 \ \Omega = 540 \ \Omega \| 562.3121 \ \Omega$$

$$R_{EQ} = \frac{540 \ \Omega \cdot 562.3121 \ \Omega}{540 \ \Omega + 562.3121 \ \Omega} = 275.4651 \ \Omega$$

5.

$$R_2 \| R_3 = \frac{0.970 \ \text{M} \cdot 0.860 \ \text{M}}{0.970 \ \text{M} + 0.880 \ \text{M}} = 0.4558 \ \text{M}$$

$$R_{EQ} = R_1 + 0.4558 \ \text{M} + R_4 = 15 \ \text{M} + 0.4558 \ \text{M} + 11 \ \text{M} = 26.4558 \ \text{M}\Omega$$

6.

$$R_3 + R_4 = 6.1 \ \text{k} + 1.3 \ \text{k} = 7.4 \ \text{k}$$

$$R_2 \| 7.4 \ \text{k} = 2.4 \ \text{k} \| 7.4 \ \text{k} = \frac{2.4 \ \text{k} \cdot 7.4 \ \text{k}}{2.4 \ \text{k} + 7.4 \ \text{k}} = 1.8122 \ \text{k}$$

$$R_{EQ} = R_1 \| 1.8122 \ \text{k} = 1.5 \ \text{k} \| 1.8122 \ \text{k}$$

$$R_{EQ} = \frac{1.5 \ \text{k} \cdot 1.8122 \ \text{k}}{1.5 \ \text{k} + 1.8122 \ \text{k}} = 0.8207 \ \text{k}\Omega$$

7.

$$R_3 + R_4 = 16 \ \Omega + 22 \ \Omega = 38 \ \Omega$$

$$R_2 \| 38 \ \Omega = \frac{47 \ \Omega \cdot 38 \ \Omega}{47 \ \Omega + 38 \ \Omega} = 21.0118 \ \Omega$$

$$R_{EQ} = R_1 \| 21.0112 \ \Omega = 15 \ \Omega \| 21.0118 \ \Omega$$

$$R_{EQ} = \frac{15 \ \Omega \cdot 21.0112 \ \Omega}{15 \ \Omega + 21.0112 \ \Omega} = 8.7520 \ \Omega$$

8.

$$R_3 \| R_4 = 4.7 \ \text{M} \| 2.2 \ \text{M} = \frac{4.7 \ \text{M} \cdot 2.2 \ \text{M}}{4.7 \ \text{M} + 2.2 \ \text{M}} = 1.4986 \ \text{M}$$

$$R_{EQ} = R_1 + R_2 \| 1.4986 \ \text{M} = \left(0.950 \ \text{M} + 1.2 \ \text{M} \right) \| 1.4986 \ \text{M}$$

$$R_{EQ} = \frac{2.15 \ \text{M} \cdot 1.4986 \ \text{M}}{2.15 \ \text{M} + 1.4986 \ \text{M}} = 0.8831 \ \text{M}\Omega$$

Multi-Resistor Calculations 1

1.

$$R_{EQ} = 1\,k + 2\,k = 3\,k\Omega$$

$$I = \frac{V_B}{R_{EQ}} = \frac{6\,V}{3\,k} = 2.0000\ mA$$

$$V_1 = IR_1 = 2.0000\ mA \cdot 1\,k = 2.0000\ V$$

2.

$$I = \frac{V_B}{R_2} = \frac{9\,V}{8\,k\Omega} = 1.1250\ mA$$

3.

$$R_2 \parallel R_3 = 10\,k \parallel 5\,k = \frac{10\,k \cdot 5\,k}{10\,k + 5\,k} = 3.3333\,k$$

$$R_{EQ} = R_1 + 3.3333\,k = 5\,k + 3.3333\,k = 8.3333\,k$$

$$I_1 = \frac{V_B}{R_{EQ}} = \frac{9\,V}{8.3333\,k} = 1.0800\ mA$$

$$V_1 = I_1 R_1 = 1.0800\ mA \cdot 5\,k = 5.4000\ V$$

$$V_3 = 9\,V - V_1 = 9\,V - 5.4000\,V = 3.6000\ V$$

$$P_{R3} = \frac{V_3^2}{R_3} = \frac{(3.6000\ V)^2}{5\,k} = 2.5920\ mW$$

4.

$$R_2 \parallel R_3 = \frac{2\,k \cdot 2\,k}{2\,k + 2\,k} = 1\,k$$

$$R_{EQ} = R_1 + 1\,k + R_4 = 2\,k + 1\,k + 4\,k = 7\,k\Omega$$

$$I_1 = \frac{V_B}{R_{EQ}} = \frac{6\,V}{7\,k} = 0.8571\ mA$$

$$V_4 = I_1 R_4 = 0.8571\ mA \cdot 4\,k = 3.4284\ V$$

Multi-Resistor Calculations 2

1.

$$R_2 \| R_3 = \frac{2.2 \text{ k} \cdot 4.5 \text{ k}}{2.2 \text{ k} + 4.5 \text{ k}} = 1.4776 \text{ k}$$

$$R_{EQ} = R_1 + 1.4776 \text{ k} = 1.5 \text{ k} + 1.4776 \text{ k} = 2.9776 \text{ k}\Omega$$

$$I_1 = \frac{V_B}{R_{EQ}} = \frac{5 \text{ V}}{2.9776 \text{ k}} = 1.6792 \text{ mA}$$

$$V_1 = I_1 R_1 = 1.6792 \text{ mA} \cdot 1.5 \text{ k} = 2.5188 \text{ V}$$

$$V_2 = V_B - V_1 = 5 \text{ V} - 2.5188 \text{ V} = 2.4812 \text{ V}$$

$$I_{R2} = \frac{V_2}{R_2} = \frac{2.4812 \text{ V}}{2.2 \text{ k}} = 1.1278 \text{ mA}$$

2.

$$R_2 + R_3 + R_4 = 0.9 \text{ k} + 1 \text{ k} + 2.1 \text{ k} = 4 \text{ k}$$

$$R_{EQ} = 4.3 \text{ k} \| 4 \text{ k} = \frac{4.3 \text{ k} \cdot 4 \text{ k}}{4.3 \text{ k} + 4 \text{ k}} = 2.0723 \text{ k}\Omega$$

$$I_1 = \frac{V_B}{R_{EQ}} = \frac{9 \text{ V}}{2.0723 \text{ k}\Omega} = 4.3430 \text{ mA}$$

$$I_{R1} = \frac{V_B}{R_1} = \frac{9 \text{ V}}{4.3 \text{ k}} = 2.0930 \text{ mA}$$

$$I_{R2} = I_1 - I_{R1} = 4.3430 \text{ mA} - 2.0930 \text{ mA} = 2.2500 \text{ mA}$$

$$V_3 = I_{R2} R_3 = 2.2500 \text{ mA} \cdot 1.0 \text{ k} = 2.2500 \text{ V}$$

3.

$$R_3 + R_4 = 3.3 \text{ k} + 4.7 \text{ k} = 8 \text{ k}$$

$$R_2 \| 8 \text{ k} = \frac{2 \text{ k} \cdot 8 \text{ k}}{2 \text{ k} + 8 \text{ k}} = 1.6 \text{ k}$$

$$R_{EQ} = R_1 + 1.6 \text{ k} = 0.5 \text{ k} + 1.6 \text{ k} = 2.1000 \text{ k}$$

$$I_1 = \frac{V_B}{R_{EQ}} = \frac{12 \text{ V}}{2.1000 \text{ k}} = 5.7143 \text{ mA}$$

$$V_1 = I_1 R_1 = 5.7143 \text{ mA} \cdot 0.5 \text{ k} = 2.8572 \text{ V}$$

$$V_2 = V_B - V_1 = 12 \text{ V} - 2.8572 \text{ V} = 9.1428 \text{ V}$$

$$I_{R2} = \frac{V_2}{R_2} = \frac{9.1428 \text{ V}}{2 \text{ k}} = 4.5714 \text{ mA}$$

$$I_{R3} = I_1 - I_{R2} = 5.7143 \text{ mA} - 4.5714 \text{ mA} = 1.1429 \text{ mA}$$

$$V_3 = I_{R3} R_3 = 1.1429 \text{ mA} \cdot 3.3 \text{ k} = 3.772 \text{ V}$$

$$V_4 = V_B - V_1 - V_3 = 12 \text{ V} - 2.8572 \text{ V} - 3.772 \text{ V} = 5.3708 \text{ V}$$

$$P_{R4} = V_4 I_{R3} = 5.3708 \text{ V} \cdot 1.1429 \text{ mA} = 6.1383 \text{ mW}$$

4.

$$R_2 + R_3 = 4.7 \text{ M} + 1.5 \text{ M} = 6.2 \text{ M}$$

$$R_1 \parallel 6.2 \text{ M} = \frac{1.5 \text{ M} \cdot 6.2 \text{ M}}{1.5 \text{ M} + 6.2 \text{ M}} = 1.2078 \text{ M}$$

$$R_{EQ} = 1.2078 \text{ M}\Omega$$

$$I_1 = \frac{V_B}{R_{EQ}} = \frac{6 \text{ V}}{1.2078 \text{ M}} = 4.9677 \text{ μA}$$

$$V_B = V_1 = 6 \text{ V}$$

$$V_1 = I_2 R_1$$

$$I_2 = \frac{V_1}{R_1} = \frac{6 \text{ V}}{1.5 \text{ M}\Omega} = 4.0000 \text{ μA}$$

$$I_1 = I_2 + I_3$$

$$I_3 = I_1 - I_2 = 4.9677 \text{ μA} - 4.0000 \text{ μA} = 0.9677 \text{ μA}$$

$$V_2 = I_3 R_2 = 0.9677 \text{ μA} \cdot 4.7 \text{ M}\Omega = 4.5482 \text{ V}$$

$$V_3 = I_3 R_3 = 0.9677 \text{ μA} \cdot 1.5 \text{ M}\Omega = 1.4516 \text{ V}$$

Multi-Resistor Calculations 3

1.

$$R_3 \parallel R_4 = 3 \text{ k} \parallel 3 \text{ k} = 1.5 \text{ k}$$

$$R_{EQ} = R_1 + R_2 + 1.5 \text{ k} = 2 \text{ k} + 2 \text{ k} + 1.5 \text{ k} = 5.5 \text{ k}\Omega$$

$$I_1 = \frac{V_B}{R_{EQ}} = \frac{5.5 \text{ V}}{5.5 \text{ k}\Omega} = 1.0000 \text{ mA}$$

$$V_1 = I_1 R_1 = 1.0000 \text{ mA} \cdot 2 \text{ k} = 2.0000 \text{ V}$$

$$V_2 = I_1 R_2 = 1.0000 \text{ mA} \cdot 2 \text{ k} = 2.0000 \text{ V}$$

$$V_3 = V_B - V_1 - V_2 = 5.5 \text{ V} - 2.0000 \text{ V} - 2.0000 \text{ V} = 1.5000 \text{ V}$$

$$I_{R3} = \frac{V_3}{R_3} = \frac{1.5 \text{ V}}{3.0 \text{ k}\Omega} = 0.5000 \text{ mA}$$

2.

$$R_2 \parallel R_3 = 0.550 \text{ k} \parallel 0.470 \text{ k} = \frac{0.550 \text{ k} \cdot 0.470 \text{ k}}{0.550 \text{ k} + 0.470 \text{ k}} = 0.2534 \text{ k}$$

$$R_{EQ} = R_1 + 0.2534 \text{ k} + R_4 = 5.5 \text{ k} + 0.2534 \text{ k} + 1.2 \text{ k} = 6.9534 \text{ k}\Omega$$

$$I_1 = \frac{V_B}{R_{EQ}} = \frac{13.7 \text{ V}}{6.9534 \text{ k}\Omega} = 1.9703 \text{ mA}$$

$$V_1 = I_1 R_1 = 1.9703 \text{ mA} \cdot 5.5 \text{ k}\Omega = 10.8367 \text{ V}$$

$$V_4 = I_1 R_4 = 1.9703 \text{ mA} \cdot 1.2 \text{ k}\Omega = 2.3644 \text{ V}$$

$$V_2 = V_B - V_1 - V_4 = 13.7 \text{ V} - 10.8367 \text{ V} - 2.3644 \text{ V} = 0.4989 \text{ V}$$

$$I_2 = \frac{V_2}{R_2} = \frac{0.4989 \text{ V}}{0.550 \text{ k}} = 0.9071 \text{ mA}$$

$$P_{R2} = V_2 I_2 = 0.4989 \text{ V} \cdot 0.9071 \text{ mA} = 0.4526 \text{ mW}$$

3.

$$R_2 \parallel R_3 + R_4 = 37 \ \Omega \parallel 45 \ \Omega + 22 \ \Omega = \frac{37 \ \Omega \cdot 67 \ \Omega}{37 \ \Omega + 67 \ \Omega} = 23.8365 \ \Omega$$

$$R_{EQ} = R_1 + 23.8365 \ \Omega = 62 \ \Omega + 23.8365 \ \Omega = 85.8365 \ \Omega$$

$$I_1 = \frac{V_B}{R_{EQ}} = \frac{6.6 \text{ V}}{85.8365 \ \Omega} = 0.0769 \text{ A}$$

$$V_1 = I_1 R_1 = 0.0769 \text{ A} \cdot 62 \ \Omega = 4.7678 \text{ V}$$

$$V_2 = V_B - V_1 = 6.6 \text{ V} - 4.7678 \text{ V} = 1.8322 \text{ V}$$

$$I_2 = \frac{V_2}{R_2} = \frac{1.8322 \text{ V}}{37 \ \Omega} = 0.0495 \text{ A}$$

$$P_{R2} = V_2 I_2 = 1.8322 \text{ V} \cdot 0.0495 \text{ A} = 0.0907 \text{ W}$$

$$I_3 = I_1 - I_2 = 0.0769 \text{ A} - 0.0495 \text{ A} = 0.0274 \text{ A}$$

$$V_3 = I_3 R_3 = 0.0274 \text{ A} \cdot 45 \ \Omega = 1.2330 \text{ V}$$

$$P_{R3} = V_3 I_3 = 1.2330 \text{ V} \cdot 0.0274 \text{ A} = 0.0338 \text{ W}$$

4.

$$R_2 + R_3 \parallel R_4 = 1.78 \text{ M} \parallel 2.2 \text{ M} = \frac{1.78 \text{ M} \cdot 2.2 \text{ M}}{1.78 \text{ M} + 2.2 \text{ M}} = 0.9839 \text{ M}\Omega$$

$$R_{EQ} = R_1 + 0.9839 \text{ M} = 1.1 \text{ M} + 0.9839 \text{ M} = 2.0839 \text{ M}\Omega$$

$$I_1 = \frac{V_B}{R_{EQ}} = \frac{11.6 \text{ V}}{2.0839 \text{ M}\Omega} = 5.5665 \text{ } \mu\text{A}$$

$$V_1 = I_1 R_1 = 5.5665 \text{ } \mu\text{A} \cdot 1.1 \text{ M}\Omega = 6.1232 \text{ V}$$

$$V_4 = V_B - V_1 = 11.6 \text{ V} - 6.1232 \text{ V} = 5.4768 \text{ V}$$

$$I_3 = \frac{V_4}{R_4} = \frac{5.4768 \text{ V}}{2.2 \text{ M}\Omega} = 2.4895 \text{ } \mu\text{A}$$

$$I_2 = I_1 - I_3 = 5.5665 \text{ } \mu\text{A} - 2.4895 \text{ } \mu\text{A} = 3.0770 \text{ } \mu\text{A}$$

$$V_2 = I_2 R_2 = 3.0770 \text{ } \mu\text{A} \cdot 0.910 \text{ M}\Omega = 2.8001 \text{ V}$$

$$P_{R2} = V_2 I_2 = 2.8001 \text{ V} \cdot 3.0770 \text{ } \mu\text{A} = 8.6159 \text{ } \mu\text{W}$$

Chapter 13

Mirror and Lens Calculations

1.

$f = -8.00$ in

Negative focal length used for convex mirror.

$$h_{object} = 4.50 \text{ ft} \cdot \frac{12 \text{ in}}{1 \text{ ft}} = 54.0 \text{ in}$$

$$d_o = 25.0 \text{ ft} \cdot \frac{12 \text{ in}}{1 \text{ ft}} = 300.0 \text{ in}$$

$d_i = ?$

$M = ?$

$$\frac{1}{d_o} + \frac{1}{d_i} = \frac{1}{f}$$

$$\frac{d_i d_o}{d_o} + \frac{d_i d_o}{d_i} = \frac{d_i d_o}{f}$$

$$d_i + d_o = d_i \cdot \frac{d_o}{f}$$

$$d_i - d_i \cdot \frac{d_o}{f} = -d_o$$

$$d_i(1 - \frac{d_o}{f}) = -d_o$$

$$d_i = \frac{-d_o}{1 - \frac{d_o}{f}} = \frac{d_o}{\frac{d_o}{f} - 1} = \frac{300.0 \text{ in}}{\frac{300.0 \text{ in}}{-8.00 \text{ in}} - 1} = -8.219 \text{ in}$$

$$M = -\frac{d_i}{d_o} = -\frac{-8.219 \text{ in}}{300.0 \text{ in}} = 0.0274$$

$$M \cdot h_{object} = 0.0274 \cdot 54.0 \text{ in} = 1.48 \text{ in}$$

2.

$$f = 6.0 \text{ m} \cdot \frac{100 \text{ cm}}{1 \text{ m}} \cdot \frac{1 \text{ in}}{2.54 \text{ cm}} = 236 \text{ in}$$

$$h_{object} = 5 \text{ ft } 6 \text{ in} = 66 \text{ in}$$

$$d_o = 2.0 \text{ m} \cdot \frac{100 \text{ cm}}{1 \text{ m}} \cdot \frac{1 \text{ in}}{2.54 \text{ cm}} = 78.7 \text{ in}$$

$$d_i = ?$$

$$M = ?$$

$$\frac{1}{d_o} + \frac{1}{d_i} = \frac{1}{f}$$

$$\frac{d_i d_o}{d_o} + \frac{d_i d_o}{d_i} = \frac{d_i d_o}{f}$$

$$d_i + d_o = d_i \cdot \frac{d_o}{f}$$

$$d_i - d_i \cdot \frac{d_o}{f} = -d_o$$

$$d_i (1 - \frac{d_o}{f}) = -d_o$$

$$d_i = \frac{-d_o}{1 - \frac{d_o}{f}} = \frac{d_o}{\frac{d_o}{f} - 1} = \frac{78.7 \text{ in}}{\frac{78.7 \text{ in}}{236 \text{ in}} - 1} = -118.1 \text{ in}$$

$$M = -\frac{-118.1 \text{ in}}{78.7 \text{ in}} = 1.50$$

$$M \cdot h_{object} = 1.50 \cdot 66 \text{ in} = 99 \text{ in}$$

$$\frac{99 \text{ in}}{12} = 8 \text{ ft } 3 \text{ in}$$

3.

$f = 250.0$ mm

$d_0 = 3{,}880$ mm

$d_i = ?$

$$\frac{1}{d_o} + \frac{1}{d_i} = \frac{1}{f}$$

$$\frac{d_i d_o}{d_o} + \frac{d_i d_o}{d_i} = \frac{d_i d_o}{f}$$

$$d_i + d_o = d_i \cdot \frac{d_o}{f}$$

$$d_i - d_i \cdot \frac{d_o}{f} = -d_o$$

$$d_i (1 - \frac{d_o}{f}) = -d_o$$

$$d_i = \frac{-d_o}{1 - \dfrac{d_o}{f}} = \frac{d_o}{\dfrac{d_o}{f} - 1} = \frac{3{,}880 \text{ mm}}{\dfrac{3{,}880 \text{ mm}}{250 \text{ mm}} - 1} = 267 \text{ mm}$$

4.

$$h_{object} = 4.10 \text{ ft} \cdot \frac{12 \text{ in}}{1 \text{ ft}} = 49.2 \text{ in}$$

$$w_{object} = 2.75 \text{ ft} \cdot \frac{12 \text{ in}}{1 \text{ ft}} = 33 \text{ in}$$

$$d_i = 267 \text{ mm} \cdot \frac{1 \text{ cm}}{10 \text{ mm}} \cdot \frac{1 \text{ in}}{2.54 \text{ cm}} = 10.512 \text{ in}$$

$$d_0 = 3{,}880 \text{ mm} \cdot \frac{1 \text{ cm}}{10 \text{ mm}} \cdot \frac{1 \text{ in}}{2.54 \text{ cm}} = 152.76 \text{ in}$$

$h_i (image) = ?$

$w_i = ?$

$$M = -\frac{d_i}{d_o} = -\frac{10.51 \text{ in}}{152.76 \text{ in}} = -0.06881$$

$$h_i = M \cdot h_{object} = -0.06881 \cdot 49.2 \text{ in} = 3.39 \text{ in}$$

$$w_i = M \cdot w_{object} = -0.06881 \cdot 33 \text{ in} = 2.27 \text{ in}$$

5.

$$d_o = 8.75 \text{ ft} \cdot \frac{12 \text{ in}}{1 \text{ ft}} = 105 \text{ in}$$

$$d_i = 15.5 \text{ in}$$

$$f = ?$$

$$\frac{1}{d_o} + \frac{1}{d_i} = \frac{1}{f}$$

$$f\left(\frac{1}{d_o} + \frac{1}{d_i}\right) = 1$$

$$f = \frac{1}{\frac{1}{d_o} + \frac{1}{d_i}} = \frac{1}{\frac{1}{105 \text{ in}} + \frac{1}{15.5 \text{ in}}} = 13.5 \text{ in}$$

6.

$$f = 28.00 \text{ mm}$$

$$d_i = 28.20 \text{ mm}$$

$$d_o = ?$$

$$\frac{1}{d_o} + \frac{1}{d_i} = \frac{1}{f}$$

$$\frac{d_o d_i}{d_o} + \frac{d_o d_i}{d_i} = \frac{d_o d_i}{f}$$

$$d_i + d_o = \frac{d_o d_i}{f}$$

$$d_o - \frac{d_o d_i}{f} = -d_i$$

$$d_o\left(1 - \frac{d_i}{f}\right) = -d_i$$

$$d_o = \frac{-d_i}{1 - \frac{d_i}{f}} = \frac{d_i}{\frac{d_i}{f} - 1} = \frac{28.20 \text{ mm}}{\frac{28.20 \text{ mm}}{28.00 \text{ mm}} - 1} = 3{,}948 \text{ mm}$$

$$d_o = 3{,}948 \text{ mm} \cdot \frac{1 \text{ m}}{1000 \text{ mm}} \cdot \frac{1 \text{ ft}}{0.3048 \text{ m}} = 12.95 \text{ ft}$$

7.

$f = 12.00$ in

$d_o = 3.00$ in

$d_i = ?$

$M = ?$

$$\frac{1}{d_o} + \frac{1}{d_i} = \frac{1}{f}$$

$$\frac{d_i d_o}{d_o} + \frac{d_i d_o}{d_i} = \frac{d_i d_o}{f}$$

$$d_i + d_o = d_i \cdot \frac{d_o}{f}$$

$$d_i - d_i \cdot \frac{d_o}{f} = -d_o$$

$$d_i(1 - \frac{d_o}{f}) = -d_o$$

$$d_i = \frac{-d_o}{1 - \frac{d_o}{f}} = \frac{d_o}{\frac{d_o}{f} - 1} = \frac{3.00 \text{ in}}{\frac{3.00 \text{ in}}{12.00 \text{ in}} - 1} = -4.00 \text{ in}$$

$$M = -\frac{d_i}{d_o} = -\frac{-4.00 \text{ in}}{3.00 \text{ in}} = 1.33$$

8.

$w_{o(object)} = 35.0$ mm

$d_i = 82.5$ ft $\cdot \dfrac{12 \text{ in}}{1 \text{ ft}} = 990$ in

$d_o = 7.50$ in

$f = ?$

$w_i = ?$

$\dfrac{1}{d_o} + \dfrac{1}{d_i} = \dfrac{1}{f}$

$\dfrac{f}{d_o} + \dfrac{f}{d_i} = 1$

$f\left(\dfrac{1}{d_o} + \dfrac{1}{d_i}\right) = 1$

$f = \dfrac{1}{\dfrac{1}{d_o} + \dfrac{1}{d_i}} = \dfrac{1}{\dfrac{1}{7.50 \text{ in}} + \dfrac{1}{990 \text{ in}}} = 7.44$ in

$M = -\dfrac{d_i}{d_o} = -\dfrac{990 \text{ in}}{7.50 \text{ in}} = -132$

$w_i = M \cdot w_o = 132 \cdot 35.0$ mm $= 4{,}620$ mm $= 4.62$ m

---

9.

$D_{i(initial)} = 0.88$ mm

$f_1 = 125$ mm

$f_2 = 650$ mm

$D_{e(xpanded)} = ?$

$M = \dfrac{f_2}{f_1} = \dfrac{650 \text{ mm}}{125 \text{ mm}} = 5.2$

$D_e = M \cdot D_i = 5.2 \cdot 0.88$ mm $= 4.6$ mm

10.

$f = 125.0$ mm

$h_i = 8.00 \text{ ft} \cdot \dfrac{0.3048 \text{ m}}{1 \text{ ft}} \cdot \dfrac{1000 \text{ mm}}{1 \text{ m}} = 2{,}438.4 \text{ mm}$

$d_i = 27.0 \text{ ft} \cdot \dfrac{0.3048 \text{ m}}{1 \text{ ft}} \cdot \dfrac{1000 \text{ mm}}{1 \text{ m}} = 8{,}229.6 \text{ mm}$

$h_o = ?$

$\dfrac{1}{d_o} + \dfrac{1}{d_i} = \dfrac{1}{f}$

$\dfrac{d_o d_i}{d_o} + \dfrac{d_o d_i}{d_i} = \dfrac{d_o d_i}{f}$

$d_i + d_o = \dfrac{d_o d_i}{f}$

$d_o - \dfrac{d_o d_i}{f} = -d_i$

$d_o(1 - \dfrac{d_i}{f}) = -d_i$

$d_o = \dfrac{-d_i}{1 - \dfrac{d_i}{f}} = \dfrac{d_i}{\dfrac{d_i}{f} - 1} = \dfrac{8{,}229.6 \text{ mm}}{\dfrac{8{,}229.6 \text{ mm}}{125.0 \text{ mm}} - 1} = 127 \text{ mm}$

$M = -\dfrac{d_i}{d_o} = \dfrac{8{,}229.6 \text{ mm}}{127 \text{ mm}} = 64.8$

$h_i = M \cdot h_o$

$h_o = \dfrac{h_i}{M} = \dfrac{2{,}438.4 \text{ mm}}{64.8} = 37.6 \text{ mm}$